KU-206-737

THESE POOR HANDS

THESE POOR HANDS

The Autobiography of a Miner
Working in South Wales

by

B. L. COOMBES

with a Foreword by
Rt. Hon. James Griffiths, C.H.

LONDON
VICTOR GOLLANCZ LTD
1974

First published 193[illegible]
Reissued 1974

ISBN 0 575 0188[illegible]

DEDICATED

TO

JOHN LEHMANN
(Editor of *New Writing*)

WHO CHEERED ME

BY PUBLISHING MY FIRST SHORT STORIES

AND WHO ENCOURAGED ME

TO WRITE THIS BOOK

Printed in Great Britain by
Lowe & Brydone (Printers) Ltd, Thetford, Norfolk

FOREWORD

by Rt. Hon. James Griffiths, C.H.

(*Former President Welsh Miners' Union*)

Those were the days when Coal was the King in Britain, and the rugged valleys of Wales the richest region in his Kingdom. For in those crowded years 1893–1913, the cry went out for Coal and still more Coal. Steam Coal to drive our trains and sail our iron ships; Ring Coal to heat our steel furnaces; and Anthracite Coal to warm our hearths. And beneath the rugged valleys which cut through South Wales there was an abundance of all these.

To get the Coal we needed the men. They came from the slopes of Snowdonia and the lands beyond the Severn to the new mining villages that rapidly developed in my native land. The men, young and old, came in search of work and wages—and that mystical something they called "life". The change from the quiet of the farm to the turmoil of the industrial valleys was a traumatic experience. The work was hard and perilous. The working hours were so long that during winter months we only saw daylight on Saturday afternoon and on Sunday. Each day as we descended into the mine we gambled with our lives. For in those early days four men were killed in the pits every twenty-four hours, and almost every year there was a terrible explosion which killed or maimed scores, or even hundreds, of our fellow workers. But work was regular, the wages better than those on the farm, and there was the attraction of living in a community throbbing with activity. There was the Workman's Club where beer and gossip helped one to forget; there was the sloping field on the hillside where we played rugger; and there were the chapels where we learnt to sing and sometimes to pray. All of which enriched the close fellowship of people who shared a common danger.

Now and again there would join the trek to the valleys some young man with the soul of a poet—such a one was Bert Coombes, the author of *These Poor Hands.* He learnt to express himself in simple direct prose which portrayed the miner's life and work in words that abide in our memory. His writings were inspired by the companionship of kindred souls. There came the day when King Coal was dethroned and the valleys were condemned to a generation of poverty and strife. And so to the long agony of the miners' struggle of 1926, which none of us who lived through it can ever forget. One of Bert Coombes' contemporaries has, in a poem, expressed what we all feel about that experience.

Idris Davies wrote:

"Do you remember 1926? That Summer of Soups and Speeches, The Sunlight on the idle wheels, and the deserted crossings. . . . Ay, Ay, we remember 1926, and we shall remember 1926 until our blood is dry."

It is appropriate that *These Poor Hands* should be read again as the wheel comes full circle. For in the 1970's as in the 1890's Coal has again become King, and the call goes out again for more and more miners.

For the young men who may answer this call the reissue of Bert Coombes' book gives a fine opportunity to read of the life of their forbears in the early days when Coal was King; and for the dwindling band of his contemporaries there is the joy of savouring again the warm fellowship of long ago.

J. G.

20 March 1974

CHAPTER ONE

I WAS fascinated by that light in the sky. Night after night I watched it reddening the shadows beyond the Brecknock Beacons, sometimes fading until it only showed faintly, then brightening until it seemed that all the country was ablaze.

The winter wind that rushed across the Herefordshire fields where the swedes rotted in heaps, and carried that smell of decay into the small farmhouse which was my home, seemed to encourage the burning, until the night sky would redden still more. Sometimes I felt sure that I could see these flames and feel their warmth, but it could only have been fancy, for they were more than sixty miles away from us.

Then in the cold and wetness of the winter evenings, when we had finished feeding the animals and had cut enough chaff for the next day, we crowded near the fire of damp logs that Mother was coaxing into flame with the bellows. I would look at our feeble fire and think, with longing, of the heat and brightness that must be about those distant flames.

We could not get a good enough price for our swedes to make it worth the six miles of cartage to the station, and the grass was spoiled in the orchard where the unwanted apples had fallen, but every night we shivered in our damp clothes because coal was too dear for us to buy. We did get some before each Christmas, because some years before a lady had left a sum of money sufficient " for eight poor families to be given one half-ton of coal each and the carter thereof to have one ton for his services ". We had the contract to be the " carter thereof ", so we had coal at Christmas time and as long after as carefulness made possible.

I can remember how I stood minding the horses while my father loaded the coal at the station, and how I pushed my hands under the horse's collar so that my fingers would keep warm. I was astounded to see several full trucks of coal and was puzzled as to how they managed to get it into a truck. I asked the porter about this mystery, and he did not seem to be any more of an expert on coal-loading than myself.

By the time I was eighteen years old I had decided that I must get away somewhere. There was plenty of work at home, but little pay. It was a very dear holding that we rented, and all ready money had to be saved for rent. New clothes were very rare, and pocket-money was something to imagine.

This did not suit my ideas of life. I wanted good clothes, money to spend, to see fresh places and faces, and—well, many things.

I had a deal of advice about my future from our two nearest neighbours. They were time-expired soldiers, and lived next door to one another about half a mile from our place. Both were bachelors, and did their own house-work—occasionally. They were often at our place, and it was usually about four o'clock in the morning when they arrived, laden with as many dead rabbits as they could carry. I have seen them bring seventy between them. They would throw them into the back of the pony-trap, and my father would get away early to town to sell them. They always called before he returned, and it was my job to give them some weak cider to soothe their thirsts until father returned and brought the money for them to have a real drink at the Comet Inn.

During one of these waits I told them of my determination to go away. Both agreed that the part of the country was " as dead as a doornail, so it be ".

" Yuh oughter join the Army same as we did "—the one named Jack Elton was definite—" an' it's in the Lancers as yuh oughter be. They'd make better'n six foot on yuh if yuh was ter join now. They w'ud so."

" Like a bolted cabbage you'd be, all length and no heart," George Jones—he was known as Tiger Jones—disagreed emphatically, as he always did; " it's in the Hussars as a lad the likes on him should be by good rights. That be the outfit as the real men joins."

Jack Elton drew his two yards and two inches to their straightest. He felt more confident that way, because Tiger was four inches shorter.

" If there's any bloke as 'ave got something disrespectful to say agen the Lancers "—Jack looked very threateningly at Tiger—" then he'd best say it when I'm not a'hearing, so he did."

" They'm a lot o' booby fighters," Tiger insisted; " they've got ter be narrer in that outfit so's they kin hide a'hint them lances. That's what them lances is for: so's they kin pop behind them if they gets into danger by mistake. This youngster kin whack me with his left lead three times out on fower, and me as spry with me hands as when I was runner-up for the championship of the Army in India. Me feet ain't so quick maybe, but I'm too good fur any scraggy Lancer, even now."

Tiger noticed my interest in the glare on the sky one morning and was sympathetic.

" If so be as you'm got no liking for the Army," he told me, " then up there in the works is the place for a young feller. Shorter hours and good money, not like as it be hereabouts—gotter graft all the hours as God sends. Ain't got to call no manner of man sir up there—no, yuh ain't.

" That be the Bessemer Works a'lighting up," he explained the light on the sky; " an' yuh could see to read in the streets of Dowlais now, so yuh could."

I had a friend, a little older than I was, who had gone to work in a colliery some months before. He was one of a family of nine who had been reared on a wage of sixteen shillings a week. His father was a good workman who started his work before five o'clock each morning and kept on at it until eight o'clock at night, caring for the horses after doing the eight hours, work on the fields. I have seen this old waggoner take off the poor socks that he was wearing and hold his feet in cold water to ease them. His feet were the colour of liver, because all the skin had been rubbed off by walking all day across ploughed ground.

At sixty years of age he had a wonderful stroke of luck, so wonderful that he could hardly believe it. He was given work on the roads at eighteen shillings a week, and would only have to work from seven to six.

I wrote to the son, Jack Preece, telling him what I thought and asked how things were going with him. He answered by return. His only complaint was loneliness, and he hoped I would come to him. He had already asked for work for me, and got good lodgings ready.

I decided to go. My parents did not like losing their only son, but they realised that things were hopeless at home, so they consented.

When I hear people extolling the joys of country living I think of the struggles of the small farmers, as I knew them. Starting work at five in the morning and leaving off just in time to go to bed about ten at night. Every footstep hampered by mud in the winter-time. I recall how my father had to work. He was alert, active, always hurrying to do something, but was handicapped by the limp caused by a horse falling on him during the South African War. He had to make long journeys to the fields he rented, for we had only small patches of ground, and the big fields and meadows near our place were farmed by bigger owners.

We paid three pounds an acre for our land, and looked over the fences at land held by big farmers for seventeen and sixpence an acre. Day and night we were afraid that the big herds of our neighbours would burst through our fences and eat up all our crop. We knew that we would get no fair recompense, yet we had to keep friendly with the wealthier farmers, because they sometimes loaned us implements we could not afford to buy.

My father once asked a gentleman farmer to rent him a piece of ground that had been allowed to become covered with hawthorn bushes and did not look to have any value. My father wanted to clear these bushes, drain and fence the meadow, and make it all look neat again—he was that sort of man. He was given a definite refusal: " Certainly not. That's the most valuable piece of ground I have got."

Some months later the same gentleman stopped my father and said, " I suppose you have heard that I am standing at the next election. We've been neighbours for some years. Can I count on your vote?"

It was not my father's way to avoid the truth. "Certainly not," he replied; "my vote is the most valuable thing I have got."

I have since wondered did he realise the truth of this statement. It has taken me several years to find out how true it is.

My mother too—like all women of her class—with her work never finished. Hunting for eggs in the barn and hedges; skimming the milk and taking all day—in the winter—to coax reluctant cream to become solid butter; helping to rear and tend animals until they learned to love them, then compelled to sell them at any sacrifice so that the rent should be paid. No holiday, week-day or Sunday, and no other prospect but to get greyer and weaker with the years, until the grave soothes with its long rest; and then even that last bit of ground having to be paid for very dearly, despite the tithe they had been forced to pay all their lives.

Yet there were others living around us who were worse off than we were—the farm labourers. They had no right to call a word their own.

Candlemas day—the second of February—is moving day in the country. It is selected because of the gardens. Every year at that date many processions went past our gate. Usually it would be a large waggon drawn by three horses and containing as much furniture as could be pushed on a handcart. These waggons would be taking the goods and families of the labourers from our locality or bringing others in their place. The women and children were covered by oilskin sheets inside the waggon, or peeping out from under the sheets as do the animals on the way to market.

Often they could not afford to visit their new home before moving, and on arrival the worried mother would find that the roof leaked; the oven was broken and the chimney smoked; that the nearest shop was two miles away along a cart-track and would not trust strangers; and that there was not a dry stick to be found to get supper with or to get early breakfast for the man and the toddlers who had to struggle through all weathers to the strange school.

They suffered most of the discomforts of gypsies, but had none of the joys which must compensate those wanderers.

I would like to feel that the old tied-cottage system is finished, but I know that things have not altered greatly. They had to spend long days at the work of a farmer who had the man and the woman—and growing daughter sometimes—under his control. He could, and often did, make them homeless and wageless for the least opposition to his wishes.

I was down in that part last year, and was told of a young married man who was instantly dismissed by a farmer for going into a public-house one night after he had finished work.

I felt like a walk around on my last evening at home, and went to watch the people going to church—it was the only bit of excitement there. It was a very large church for the small parish and had a fine set of bells. I noted that the ringers seemed to go very easy until the carriage-and-pair of the squire rounded the turn by the old preaching-cross and the great man began to struggle out of his carriage. Then they pealed in earnest. Every worshipper had to wait outside until the squire had walked to the widening of the path and had made that dramatic flourish when he pulled out his gold hunter watch and looked up at the church clock. When he was satisfied that the clock had not dared to contradict the time shown on his watch he would nod to the clock, smile at the admiring people, and hold out his hand to the vicar standing in the doorway to welcome him.

Then the bells would ring merrily and from the other direction the staff of another big house marched to the church: housekeeper and butler in front, two footmen next, then about fourteen girls walking in pairs. They were paraded to church every Sunday, but were only allowed one free evening a month. This rule was considered very harsh by the young men of our village and probably by the girls themselves.

I did not relish being one of the few worshippers in that large church on such a fine evening, so I walked away until I came to the main roadway. It was well named—Stony Street. At the lower end it joined the Hereford Road, and near this junction was the Comet Inn. Outside the window of the inn Tiger Jones was sitting, with his pint mug on top of a tree-trunk that had been sawn to table height.

Knowing Tiger so well, I guessed what had happened, and very soon found I was correct. He had not been at all Sabbath-like in his language, and the landlord had wisely gone from sight and left his wife to order Tiger outside. So outside he was, looking as miserable as a disgraced schoolboy, and was dodging the decision that he should have no more beer on those premises that night by getting Elton to pass it through the opened window when no one was looking. It seemed, though, that Elton was too interested in the gossip inside to attend properly to the thirst of Tiger outside.

"That there Elton," Tiger greeted me, "he be quite as much a scoundrel as that there Porter as calls himself the landlord. Mark my words, young 'un, an I'll be doing the lot on 'em some depredation afore this day be gone, I will that."

"Have a drink with me?" I asked the unnecessary question.

"I'll be everlasting obliged to yuh," Tiger agreed, "if so be as yuh can get it out ter me without that old witch of a Mother Porter a'seeing of yuh."

When I returned with the full pint I paused in the shadow of the porch, for another man was approaching Tiger. This was an evangelist on his way to give a service at a local chapel, and he saw in Tiger a possible convert. Tiger looked up when the other came near, and that look checked the greeting of the newcomer. He felt in his pocket hurriedly, found a tract, and handed it to Tiger, who studied it as an owl might study a notice that stated "Trespassers will be prosecuted".

"Well,"—Tiger spoke first—"an' what might you reckon to call this 'ere, hey?"

"My friend,"—the evangelist was nervous under Tiger's glare—"that's for you. It's a letter, you see. Yes, a letter from God."

"Well, I'll be damned." Tiger studied the printing with interest. "It's a heck of a time since I heard from him, so it is."

After the evangelist had hurried away, uncertain whether he could count Tiger as a victory or a defeat, I took out the drink.

"It'll be a farewell drink," I explained, "I'm off to-morrow."

"Eh? What's that?" Tiger was so surprised that he held the mug in his hand and did not start to drink. "What d'you mean?"

" I'm off to-morrow," I repeated, trying to be casual.

" To-morrow? " he repeated. " Are you now? Well, I'm proper sorry, I am. Yet it'll be nice to have some one from here in the old crush. It's a credit to 'em you'll be."

" No, thanks," I said, " no army for me. I'm going up to Jack."

He took one drink from the mug and left enough to cover the bottom. He handed the mug to me and insisted that I finish it off.

" We've allus been pals, you an' me," he said slowly, " and right to the end we shares. Wish as I wus a bit younger so's I could come with yuh. Look after yourself, and whatever yuh does keep away from them Welsh gels. Lord! They do say as some on 'em is holy terrors, so they be."

Next morning we lifted my tin box on the two-horsed carrier's brake and started for Hereford. I hated leave-taking, and made it as brief as possible; besides, it would not be so long before I returned on holiday with good clothes and money to spare.

I looked back from the turn at the old home and heard the dog barking his sorrow at the parting. I carried the smell of the wood fire with me, and it has hung in my senses ever since. Little things like the thud of a falling apple, the crackle of corn being handled, the smell of manure drying on a warm day, the hoot of the owl from the orchard at night, and the smell of the new bread when my mother drew it from the stone oven on a long wooden ladle, are still very sweet to me. Every year the smell of drying grass makes me crave for the hayfields, but I have never since worked in them or been in my native place for anything more than a short visit.

We travelled slowly townwards, and had frequent stoppages while housewives gave detailed instructions to our driver about goods they wanted from town and exchanged gossip and greetings to some of the other passengers.

At the few slopes—they were hardly hills—the most active of the passengers lightened the load by getting out and walking. It was nearly an hour and a half later when we rattled down the slope past Belmont Monastery and could see the spires of Hereford gleaming into the blue sky.

That road slopes downwards into the city until it crosses the old bridge over the Wye. Most of the passengers got off near the Cathedral, but I was driven through Broad Street and under the large shadow of All Saints' Church, through the High Town, past the black-and-white pile of history that is called the Old House and now contains many reminders of the days of Nell Gwynne, Kemble, and Garrick. It was not market day, so there were no droves of white-faced Hereford cattle in the streets, no Ryeland sheep bleating their way along; only the sour smell of the Hop Market and when we passed the Corn Exchange the mixed smells that reminded me of harvest.

There can be few more peaceful, and beautiful, cities in the world than Hereford, and at the last I did not like leaving it. If one could only have a job that gave one enough to live upon, I thought, then life could be very good in this pleasant spot; but it seems one of the ironies of life that the worker must always go away from such spots before he can get a decent wage.

CHAPTER TWO

THE carrier's brake turned away from the station and left me and my tin box to the world. No one took any notice of me—the railway-men carried on with their tasks, the farm-hands worked in the fields, all unconcerned with the event that made me feel so alone and miserable.

I had labelled the tin box very plainly, and meant to be near my clothes whatever happened. I was not more than two doors away from the luggage-van during the whole journey. I watched that luggage door at every stop; if the box was lifted out I was soon alongside it. When it was placed in another train I got in two doors away.

I thought the Crumlin Bridge a wonderful thing, and under it was the first real valley I had ever seen. The day had become greyer and the countryside was no longer green and flat. Grey streets clung to the hillsides and seemed only just able to avoid slipping down into the river in the bottom of the narrow valleys.

The train was more crowded and the talk was cheerful and friendly. Most of the men wore bowler hats as if they were compelled to, and their clothes were a contrast to whiter faces than I had ever seen. The women were mostly short in build and dark. They were very animated.

At Quaker's Yard I was left on my own. The box was not disturbed, and we rolled slowly onward. It was beginning to get dark. About ten minutes later the train screamed into a tunnel, and when we came again to daylight I found that we had left the greyness of the narrow valleys behind. Here was beauty of scene once again, but not the flat smoothness of the English Midlands; instead I saw the opening of a wide valley shut in by splendid mountains. On the left side of the railway-line the sides were solid rock that mounted higher and still higher until I had to lean out over the window to see the top.

Below, to the right, lights showed among the trees. The distance below and the thickness of green foliage made it all seem fairy-like. Over more to the right I could see the glare of an immense fire lighting up the evening sky. I was getting very close to the light on the sky I had watched so much, for only the width of a mountain was between that fire and me.

Then painted on the station board I noticed the name I had been repeating all day—we will call it Treclewyd—and the train shuddered at the harshness of the braking. The station was crowded with people who seemed to be going to the market town for the night. They all crowded into the train as if it were the only one that ever travelled that way, and I had a job to find out whether the precious box was unloaded.

I had noticed that more Welsh than English had been spoken among the crowd who got on that train. I found myself alone with a young porter, who first studied my labels, then became interested in me.

" Going to stop here? " he inquired.

I admitted it.

" I'm from Herefordshire too," he stated, " and I'm after a change back there as soon as they'll give it me."

I did not reply to this, as I was too busy thinking about the advantage of working on a job that gave such desirable facilities. Then he added, in the tone of one who offers a secret way of escape:—

"There's only one train that way on Sundays, but there are quite a few on other days."

I found out afterwards that I had arrived earlier than I had expected, so Jack was not there to meet me. The porter decided to help a fellow-exile, and assisted me to carry the box up the gravel path to the street. We had not gone more than a hundred yards when he lowered his end and stated:—

"That's the house. See you some time agen. So long."

A lot of elbow-grease must have been used to cause the polish on that door. Despite the rain which was falling, the brass rod at the bottom shone and the step was milk-white.

A girl of about my own age opened the door, and looked in astonishment at the box and me. After some explanation I was taken inside, and the two daughters as well as the father and mother helped to deposit the battered box near the bottom of the stairs, where it contrasted with the shining passage, the bright stair-carpet, and the brass rod on each step. The kitchen was a'dazzle with brass. There was a row of brass candlesticks on the mantel and a strip of brass along the edge, as well as a thick brass rod beneath and another wide strip covering the upper part of the chimney-opening. That fire did not peep from under the second bar, it filled the grate as high as was safe, and its white heat showed in the reflections on the fender.

The warmth and comfort were something to which I had not been used. In the country we did not make much use of the house. We went in muddy to meals and hurried out to work soon after. We only stayed for any time when we were in bed, for the stables were nearly as warm and comfortable, so I was astonished at the clean comfort of this new abode.

I have found most Welsh mining-houses as clean—or nearly so—as this one. The women work very hard—too hard—trying to cheat the greyness that is outside by a clean and cheerful show within. They age themselves before they should because of this continual cleaning and polishing.

After tea my new landlord took me to look for Jack's lodgings. I saw only a very few real shops in the village—most of the selling was done in houses that had converted their front rooms into shops and hung outside signs advertising tea or salmon. The main street was a crooked affair, it twisted about every few yards, and the side streets came into it abruptly, with the end of the top house jutting out into the main street. Most of these house-ends were covered with advertisements.

Jack lodged in the third of these side streets. One feeble lamp was placed at the junction of the main road, and after that there was no more light for nearly a hundred houses, all of which were built of the same kind of dark stone and had the doors and windows painted the same colour. Big stones jutted up in the roadway and smaller ones made the side-walk difficult.

A hawker was leading his pony carefully between the stones and loudly proclaiming that his apples were cheap at eightpence a pound. The customers stumbled out to the cart and groped their way back from the faint light of the trap lamps.

" Yes, I expect there are plenty of apples where

you come from, isn't there?" My landlord asked the question between stumbles.

"Yes, too many," I replied. "They offered us seven bob a hundredweight for best pot-fruit, better than them"—I knew the prices off pat—"and about thirty bob a ton for windfalls for jam. Windfalls are those that have fallen off the tree, you know."

"Off trees?" he answered, and was obviously puzzled, whether at the method of growing apples or at the price I never knew—"Falling off the trees? Well, I declare."

"I wonder what he would charge for those swedes?" I inquired.

"Swedes?"—the landlord seemed proud of his knowledge"—Well, indeed. Swedes are very cheap—yes, they are. Can get them for a penny a pound."

He took hold of my arm, partly not to lose me in the darkness and partly in case we should fall, I suppose, and when he tried to discover exactly where we were, I reflected on the tremendous difference that the sixty or seventy miles of my journey had made in the value of apples and swedes. A week before we had been offered twelve shillings a ton for swedes when placed in the truck.

My landlord began to think he was near the right door, so he knocked—and was only two doors out. He was pleased not to have been farther away. We had a good welcome at Jack's lodgings, and were reproved for not "walking in without knocking". This seemed to be the custom, for all that evening someone or other was walking straight in. There was none of the "keep away from this place, I pay rent for it" about these people, as is the case in the English farming areas. These people were all friendly, they came in and joined in the talk. Most of the visitors seemed to be females who came to look at this newcomer and listen to his strange accents on the few occasions when he could be persuaded to talk.

Jack came in with his shopping soon after. He stepped on a mat in the passage and slid gracefully into the room. He held on to his goods very well, only dropping one loaf, which rolled under the grate. He partly slid under the table, but there was enough of him outside for me to see the pleased grin on his face when he noticed me. I asked him how he told one house from another and what would have happened if he had gone into the wrong house by mistake. He had done so many times, so he said, but of course not in that hurried manner.

CHAPTER THREE

My bedroom invited sleep that night, and I did my share. That household was very sedate on the Sunday morning. I caused a sensation by saying "Good heavens" during a talk. The landlady was so reproachful that I was really moderate in my language afterwards.

After breakfast I again upset things by wanting water to shave. It seemed that shaving was considered to be work in that household. It might have been so for the landlord, but in my case it was a luxury and quite unnecessary.

The eldest daughter stayed at home to prepare dinner while the rest went to morning chapel. They wanted me to go with them, but I made the excuse that I was a church-goer. I had found out that the church was in the opposite direction to the chapel. This church seemed tiny in comparison with the cathedral-like buildings down home. Few people were about there, and as no one asked me inside, I wandered around looking at the graves. In view of the danger of my new occupation, I felt a deal of satisfaction, because if anything happened to me I would not be buried in that rocky graveyard, but would be taken to the more mellow one at home.

Half an hour afterwards I returned to my lodgings and startled the daughter just when she was entranced with one of the chapters in a novel.

"My goodness!" she gasped, and she had been too surprised to hide the book. "Why on earth didn't you make more noise, you cuckoo? I thought it was father come back—I did really. But there, it's lucky you came, or I might have let the dinner burn. That would have been awkward, wouldn't it, with the minister coming to dinner?"

"The minister? Coming here to dinner?" I was alarmed.

"Nuisance, isn't it?" she agreed; "but I suppose he can't be expected to bring a bit in his pocket. Keep pushing the cabbage down with the fork when it boils over, will you, please? Would you like to have a look at this later?"

"I would that," I said, and did as required to the cabbage, for I have always been quite as willing to assist with the cooking as I have been to eat it, and that is saying a lot.

We talked over our mutual fondness for reading, and I was shown the hiding-place behind the coal-house where the books were placed. Before we thought it necessary the others returned and brought the minister.

I dodged Sunday School by going to find Jack at his lodgings. Things were peaceful there, for they were digesting their dinner and were barely energetic enough to turn over the pages of the Sunday paper. I had a deal of coaxing to do before they agreed to come for a walk and show me the sights.

Jack's landlord must have been about fifty years old. He was short and squarely built—a typical "collier", and the right size, for a tall man is often at a disadvantage underground. His nose was broad and more blue-black than any other colour, because a falling stone had smashed it into his face some years before. Every inch of his face had blue spots showing where a shot had exploded almost in front of his eyes and blown the bits of coal into the skin. His sight was good, although his eyelids and near the pupils of his eyes were dotted with these marks.

I believe that Jack had been over-stating my boxing ability to him, for he was very interested in me and said he was sorry he hadn't had me to lodge at his place. He was from the Rhondda, and gave me details of the boxers he had seen there. His conversation was either of boxers or colliery work—nothing else. After a while he thought it best to give me some tips about the work.

"There's a lot to learn, see, bachgen," he explained, "and if you don't learn it properly you'll most likely get a stone on top of you, aye indeed.

"There's not two collieries as is the same," he went on explaining; "they all got some different drawbacks, aye they have. Mostly it's the top. When it's clod above you, it's sly, and you wants to put a post every yard. Then when it's rock top they thinks as it's safer, and they wants you to go a long way without a post. Better to work in bad top, I do say; then a chap is always careful. There's cliff top, too. Dangerous, that is. It do sound strong and then it do give without a crack, aye it do."

I listened gladly and treasured his words. I fancied I could tell what things looked like underground without seeing the place, but I found myself altogether wrong later.

"There's gas, too," he continued, "although there ain't any in the level where you will be working. In the big seams it's there without you a'knowing it. Gets in your guts and makes your head spin, it do. And if you was to hit a rail with your mandril, well, up you goes, p'raps."

" You mean explosions ? " I wanted things on a big scale. " Ever been in one ? "

" Yes, a couple of small ones," he replied. " We did have a small one near home. Seven killed."

" What caused that one ? " I was eager for information.

" God knows," he answered slowly, " but the men did get all the blame, as usual. I was one of the first in there after it happened, aye I was."

" Did you help to find them ? " I was impatient for full details.

" Aye, five of them. Then the manager told us not to go into the other stall-road. He'd been there, and two of the men were inside. But we did go on the sly, and the dead men had only one lamp between them. When the inspector came he found two lamps there, one opened, and they counted as most likely they'd been smoking. Dead men can't argue—see, bachgen——and ever since when I hears of matches been found or opened lamps after an explosion I wonders how they did get there, aye I do."

In Treclewyd that afternoon dust, black dust, blew in through every open window; the many empty tins that were about filled rapidly with dust that contained a large percentage of the mixture that caused silicosis; grey house crouched tightly to grey house with no relief of colour and no garden or back entrance; and leaves were grey, and never green, and berries were black before they were red. The dread of consumption was among the people like an ever-present shadow.

Yet, away from the dirt of the village one had only to walk a few hundred yards and there was a wonderful valley sheltered by mountains that made me gasp with amazement. Trees covered their lower slopes and showed their darker greenness up to where the mountain grass and brown ferns seemed to meet the sky at the summit. The river curved across the width of the valley and coursed slowly along the ten miles that separated us from the sea at Swansea. The sun was warm in a blue sky, and we heard the song of many birds before we had gone from the sound of the Sunday-School singing. It was a grand day in a grand setting.

But because Nature had been so kind below ground as it had been above, and had placed seams of coal from eighteen inches to eighteen feet there, as well as beds of fire-clay and iron ore, man had come along and defiled the beauty and blackened with his muddling what had been formed so beautifully.

A first-class road went right along the valley, and many people were out walking. Groups of girls, dark-eyed and animated, sauntered along arm in arm in search of mischief or adventure. The young men were well dressed and keen in appearance. The family groups were more soberly dressed than the young people, for when one has many responsibilities to meet it is wise to limit the clothing to dark colours, so that when the sad days come one is prepared—and those days come suddenly to mining families.

It was one of the boom periods of the coalfield, and every one seemed contented and happy. It was a village that always maintained a good standard of dressing.

I persuaded Jack to come to the chapel service that evening. He was dubious, because he had heard of an Englishman who had gone to one of the chapels there not knowing that the service would be in Welsh. He sat through that service in complete confusion, not knowing when to get up or sit down and not understanding one word of what was said.

We did enjoy that service, however; it was in English and the singing was good. I was curious about the queer design of that chapel; the front windows were very wide and placed at a strange angle. I was told that the colliery company had agreed to lease the ground to build this chapel only on the condition that it should be designed in such a way that it could be easily altered into an engine-house if they should ever wish to do so.

On Monday morning I was to "sign on" in readiness for starting work that night. I had been coached well in what I was to say, because it had been said that I had done a little colliery work before. Otherwise I might not have been given work.

This signing on was a serious affair. I had to satisfy five different clerks and provide material for detailed entries in their books.

The first wanted assurance that I had not suffered from nystagmus, miner's chest complaints, "beat" knees or elbows, had any serious injury, or been paid compensation due to or arising from the afore-mentioned. I struggled through that lot satisfactorily.

The next clerk—he was elderly and seemed to have quarrelled with his wife that morning—asked me in a deep, sad voice for the address of my nearest relative so that they could be communicated with in case of accident or death.

The next one was no tonic either. He wanted me to sign an agreement to have a lot of my pay deducted. State insurance, hospital, doctor, district nurse, library, artificial-limb fund, blind institute, band, and a few more all came in for their coppers, which meant quite a considerable total.

I had been prepared for the next one. He wanted to know where I had worked underground previously. I mumbled something about Mountain Ash and the Dyffryn Colliery. He checked his pen, stretched his leg out, and looked straight at me. I felt that it was all over, but he asked, quite casually:—

" They call it the Deep Dyffryn, don't they? "

I nodded, feeling that if he only knew a bit about that colliery he knew more than I did, so I would not argue. He smiled at me—why, I have never been sure—and I tried to return it as I moved across to the next desk.

This clerk muttered something about agreements and contracts, then pushed a book forward for me to sign. Next he handed me a slip to show I had fulfilled all requirements and was signed on as a collier's helper. I was outside as soon as I could feel my way along the dark passage, for those offices were low and dirty—an old farmhouse roughly adapted and with its past whiteness almost covered by the small coal that was everywhere.

I had not made the mistake that another countryman made when signing at the same office. He spoke about the colliery in the Rhondda where he had worked, and the astute clerks asked him what sort of lamps were in use there.

" Lamps? " was the surprised answer. " Well, I—I don't know. You see, I never worked there in the dark."

Of course, he never started.

That afternoon I met my " butty ". He was a Welshman past middle age, quiet and sincere. His name was John——. He was an experienced miner and from the start did his best to pass his knowledge on to me. He was anxious to do right by everyone; I never saw him do a mean action. In my value of men I count him a gentleman. A man such as he would not be kept in the mines to-day, because he was too careful, too experienced, to rush his work and take the risks men are forced to do in the mines now.

About nine-thirty that night I started to dress for my first night underground. There are no rules as to what you shall wear, only an unwritten one that you must not bring good clothes unless you do not mind being teased about what you are going to do for Sunday or " how's it looking for the old 'uns? " Clothes must be tough and not too tight; dirtiness is no bar, because they will soon be much dirtier than they have ever been before. The usual wear is a cloth cap, old scarf, worn jacket and waistcoat, old stockings, flannel shirt, singlet, and pants. Thick moleskin trousers must be worn to bear the strain of kneeling and dragging along the ground, and strong boots are needed because of the sharp stones in the roadways and the other stones that fall. Food must be protected by a tin box, for the rats are hungry and daring; also plenty of tea or water is necessary to replace the sweat that is lost.

John was waiting for me near the colliery screens at ten o'clock. I was silent, and on edge for what the night would teach me. We climbed uphill between a double set of narrow rails, and the woods shut us in on either side. It took us half an hour of steady climbing before we halted on top of a grey pile that was the rubbish-tip and we could look back on the lights in the village below. The fire from the steelworks was like a red moonlight that night; we could see the expression on one another's faces by it. I was very near to my light, but did not get much comfort from that thought.

There were about fifty men sitting in the dark near the mouth of the level. We went inside a rough cabin that had closely printed extracts from the Mines Act and the Explosives Order nailed near the door. A youngish man was sitting on a box inside. I noticed that he was holding some pencilled notes to the left side of his face so that he could read them, and found out later that a piece of stone had knocked out his right eye some months before. He was the fireman, in charge of that shift because the colliery was not large enough to employ a night overman.

He seemed decent and intelligent. We handed in the "starting-slip" to explain my presence, and he told John that he had found a strong-looking mate, and told me that I had got a good butty—one of the best in the valley. I am convinced the latter statement was correct.

At five minutes to eleven we started to walk inside. Each of us carried a "naked light" or oil lamp that had no glass or gauze to restrict the light, because there was no danger of gas in that level.

As soon as we entered under the mountain I was aware of the damp atmosphere. Black, oily water was flowing continually along the roadway and out to the tip. It was up to the height of a man's knees, and to avoid it we had to balance carefully and walk along the narrow rails. I slipped several times, and then tried crouching up on the side and swinging myself along by the timber that was placed upright on either side. Suddenly I remembered that this timber was supposed to be holding the roof up and that I might pull it out of place and bring the mountain down on to us. I did not touch the timber after that.

We were not more than ten minutes reaching the coal-face—that is the name given to the exact part where coal is being cut. This was a new level, so it had not gone far into the mountain. This seam was a small one, not a yard thick, and was a mixture of steam- and house-coal.

When I had been shown where to hang my clothes I went to see our working place. It was known as the Deep. We were the lowest place of all, because this Deep was heading into the virgin coal to open work. Every fifty yards on each side of our heading other headings opened left and right, but they would be working across the slope, and so were running level. From these level headings the stalls were opening to work all the coal off.

Our place was going continually downhill. Every three yards forward took us downward another yard. It was heavy climbing to go back, and every shovelful of coal or stone had to be thrown uphill. Water was running down the roadway to us and an electric pump was gurgling away on our right side. We were always working in about six inches of water, and if the pump stopped or choked for ten minutes the coal was covered with water. There is nothing pleasant about water underground. It looks so black and sinister. It makes every move uncomfortable and every stroke with the mandril splashes the water about your body.

It takes some time to be able to tell coal from the stone that is in layers above and below it. Everything is black, only the coal is a more shining black and the stone is greyer. It is difficult to tell one from the other, especially when water is about, but the penalty for putting stone—miners call it "muck"—into a coal tram is severe.

I tried very hard to be useful that night, but was not successful, nor do I believe that any beginner ever has been. Things are so different and there is so much to learn. For several weeks lads of nowhere near my size and strength could make me look foolish when it came to doing the work they had been brought up in. I had used a shovel before, but found that skill was needed to force its round nose under a pile of rough stones on the uneven bottom, turn in that narrow space, and throw the shovelful some distance and to the exact inch.

The need to watch where you step, the difficulty of breathing in the confined space, the necessity to watch how high you move your head, and the trouble of seeing under these strange conditions are all confusing until one has learned to do them automatically. It takes a while to learn that you must first take a light to a thing before you can find it. I started several times to fetch tools, then found myself in the solid darkness and had to return to get my lamp.

My mate lay on his side and cut under the coal. It took me weeks to learn the way of swinging elbows and twisting wrists without moving my shoulders. This holing under the coal was deadly monotonous work. We—or rather my mate—had to chip the solid coal away fraction by fraction until we had a groove under it of an inch, then six inches, then a foot. Then we threw water in the groove and moved along to a fresh place while the water softened where we had worked.

John hammered continually for nearly three hours at the bottom of the coal. He cut under it until he was reaching the full length of his arms and the pick-handle. At last he slid back and sat on his heels while he sounded the front of the coal with the mandril blade and looked closely at where the coal touched the roof to see if there showed the least sign of a parting.

"Keep away from this slip," he warned me as he moved farther along, "it'll be falling just now."

It did, in less than five minutes, and after I had recovered from my alarm and most of the dust had passed I did my best to throw the coal into the tram. I soon found that a different kind of strength was needed than the one I had developed. My legs became cramped, my arms ached, and the back of my hands had the skin rubbed off by pressing my knee against them to force the shovel under the coal. The dust compelled me to cough and sneeze, while it collected inside my eyes and made them burn and feel sore. My skin was smarting because of the dust and flying bits of coal. The end of that eight hours was very soon my fondest wish.

After working for a while John went away to search for a post. About that time one of the hauliers—I never heard him called anything but Will Nosey—decided to go and see how this new starter was getting on. I was alone, and pretty nervous, when he arrived. His nickname came partly from his interest in the concerns of other people and partly because of his long nose, which curved downward as if it meant to get inside his mouth.

Our lighting was feeble, and his face showed a grey white in colour. His voice was always loud, and the hollow passage amplified it. He had a different type of lamp from mine. His was one with an open wick and it was worn on his cap. The ventilation was good, so that the flame blew back over his head. I had reason to be alarmed, for his eyes were sunken alongside that hooked nose and there was a queer grin on his grey features, while above his head the oil flame hissed and crackled. He was exactly like one of those pictures we see of the Devil.

I did not speak, but stood and stared. He did exactly the same thing. Then someone not far away fired a shot to blow the roof down. In that confined space the noise and vibration were terrific. The whole mountain seemed to shake with the powder-charge. I could not hear for some seconds after.

I had never before heard such a crash, nor had I had any warning to prepare me. I could only imagine that an explosion must have happened and that this other being with flame over his head was there to capture me.

Then the haulier, who had finished his scrutiny, and had been delighted to see how startled I had been, started to laugh. It was intended for laughter, but it sounded most sinister to me.

" He, he," he chuckled, and when I realised that he used earthly words I became more at ease. " He, he. Made you jump, did it? That was a shot for ripping top, that was. Oughter have warned us, they did; but I s'pose they was on the watch. Don't half shake things up, do it? Like to be here, hey? "

I had still enough determination left to say the lie that I did like being there. I was discovering still another new pain. My knees, unused to the hard rubbing against the stone bottom, had become like those of a Yorkshire miner whom I met later and who insisted that his knees had become " b—— red hot ".

As we were anxious to open work, we only cut the width of the roadway, about nine feet. The top coal was about one foot thick. We holed that off, then wedged up the lower coal. Altogether we filled three trams of coal, and the heading had gone forward about one yard. Then John rammed a bore-hole with powder and they lit the fuse. We went away to have food while the smoke from the shot was clearing.

We had a quarter of an hour for food. For the first time that I could remember I had no appetite, and the rats that ran about outside the circle of our lights had my food and squealed a lot while eating it.

The shot had ripped the top down, and we had to clear the stones so that the tram could be brought closer. Before that shot was fired we did not have more than a yard of height at the coal-face, and as I was clumsy I rubbed my back against the slime of the roof. My shirt-back was soon covered with a thick coat of clay and my back was getting as sore as my knees.

By four o'clock in the morning the shovel felt to be quite a hundredweight and I winced every time I touched my knees or back against anything. I got sleepy too, and felt myself swaying forward on my feet. I dropped some water in my eyes and revived for awhile, then I pinched my finger between two stones and was wide awake for some time after.

I had thought that night and day were alike underground, but it was not so. It is always dark, but Nature cannot be deceived, and when the time is night man craves for sleep. When the morning comes to the outside world he revives again, as I did.

Even the earth sleeps in the night and wakens with perceptible movement about two o'clock in the morning. With its waking shudders it dislodges all stones that are loose in the workings. It is about that time that most falls occur, at the time when man's energy is at its lowest.

Somehow that shift did end, although I felt it lasted the time of two. Then we had to lock our tools for the day. Holes are bored in the handles of the tools and they are pushed on a thin steel bar with a locking-clip fitted in the end.

John had a pile of tools, and they were all needed. Shovels, mandrils of different sizes, prising-bars, hatchet, powder-tin and coal-boxes, boring-machine and drills and several other things. He valued them at eight pounds worth, and he was forced to buy them himself. He knew they might be buried by a fall any day and was not hopeful of getting any compensation for them. Nearly every week he had to buy a new handle of some sort and fit it into the tool at his home, so that his wages were not all clear benefit, and his work not always finished when he left the colliery.

How glad I was to drag my aching body toward that circle of daylight! I had sore knees and was wet from the waist down. The back of my right hand was raw and my back felt the same. My eyes were half closed because of the dust and my head was aching where I had hit it against the top, but I had been eight hours in a strange, new world.

The outside world had slept while we worked, and the dew of the morning sparkled from a thousand leaves when I looked down on the valley. It was beautiful after that wet blackness to see the sun, and the brown mountain, and the picture that the church tower made peeping out over the trees.

As we went down the incline, the day-shift came up. They called their " Good morning " or " Shu mai? " as they hurried past with their tea-jacks in their hands.

My landlord worked nights that week, as well, and they were all ready at the lodgings. We drank a cup of warm tea by holding the handle with a piece of paper so that we should not dirty it, then prepared to bath. We used a wooden tub, made by sawing one of the company's oil-casks in half.

My landlord gave me the benefit of his experience in getting all the dirt out of my eyes and ears. This was not so easy a task. After some days my skin seemed to get used to the dust and washing was easier after. Nowadays if anything interrupts the routine of a daily bath I begin to feel scruffy and to itch all over.

The landlady warned us against splashing, then left us to it. We washed to the waist, and he washed my back and I did likewise to him. We threw towels over our shoulders, then carried the tub out to empty it at the drain. Then filled the tub with fresh water that had been warming and stripped ourselves to complete our washing. After we called out that we were dressed, the rest of the family returned and we had breakfast.

I forgot my shyness over bathing in a few days, but it was not so with Jack. He did not wash under the same conditions, either. They were much more crowded at his lodgings than we were. Another family occupied the front room there—as was the case in almost every house in that street—and Jack was miserable because some of the women came into the room while he was washing, and stayed. It was torture to Jack to wash all over in the sight of anyone, but the people there were used to that sort of thing from childhood and thought nothing of it. No doubt they noticed Jack's shyness, and had a bit of fun over it.

I knew a Welshman who went to work in the Yorkshire mines and stayed there some years. When he returned to South Wales he brought the new accent with him and settled in a fresh place. Everyone thought he was an Englishman. The women where he lodged and some female neighbours always stayed indoors while he bathed, and amongst sly glances they had a lot to say concerning his physical capabilities and shortcomings. They spoke freely to one another in Welsh and became more personal as the days passed and he showed no sign of understanding. One day they annoyed him too much and he started to abuse them in fluent Welsh. There was consternation in that circle after that and a definite silence until he found other lodgings.

It was quite half an hour after my arrival home before I was properly clean and ready for food. This time used to get rid of dirt has always seemed to me to be part of the working day. Apart from that we had more than a half-hour's walk to work each day, so that I was away from home about nine and a half hours.

Life is complicated when one works by night. I went to work one day and returned the next. We started work on a Monday night, and one hour later it was Tuesday morning and the date was one day later. I had to watch this question of dates very carefully years afterwards when I had to report a lot of accidents, as the wrong date might lose a compensation claim.

Then again, I went to bed on the Tuesday morning, stayed there for eight hours, and it was still Tuesday. It seemed strange that a night had not passed while I was in bed. There were no roaring loud-speakers in those days, but there were children playing in the street and hawkers with voices that sounded to be shaking the bed I tossed about on. The wind was the right way that morning, so that the dust of the colliery screens blew away from the village. So all the women were busy doing their washing. Tubs were bumped and washing-boards banged right underneath my window. All the water had to be carried from taps which were set at various spots throughout the village and a crowd of women were usually near these taps waiting for their water-vessels to be filled. These taps were great places for gossip, and every now and then I would hear a sharp voice calling, " Our Lizzie Ann! D'you know I'm waiting for that water?" or " Gwennie! Gwennie! You've been up at that tap all the morning."

When the wind was the wrong way I had the choice of stewing in a bedroom with a tightly closed window or swallowing coal dust while I slept on bedclothes that became blacker every minute.

That first day I dozed and awoke until I could stay there no longer—and got up. My head ached as badly as the rest of my body and my mind decided that one week would be six days too long at that job.

Jack consoled me by saying that the aching of my body would get better in a few days but that there was little hope that I would ever sleep properly by day. He was quite right, and even to-day I notice that the beginning of each night shift shows the men to be weary and disheartened by the lack of proper rest and that all are hoping the week-end will come soon so that they can have a real night's sleep.

I remember an old Welshman who detested night work as much as I did. He came to me one day and told me that " that undermanager, mun, he did ask me to work nights for a couple of day, aye he did."

" What did you tell him? " I asked, and tried not to smile at his confusion.

" That'll be the day, I did tell him," he replied, " that'll be the day—when I do work nights."

An owl hooted at us when we walked through the woods to work on the third night. Whether it was in sympathy or derision I have never been sure. I thought so much about that combination of third night and hooting owl that I would have turned back gladly, only I was ashamed to do so.

CHAPTER FOUR

By the end of the first week most of my stiffness had gone, but a deal of the soreness remained in my hands and knees. There have been very few weeks since that I have not been glad of the Sunday rest to give some cuts or bruises a chance to heal. It seems part of the job, that getting battered about.

I had a week's pay in the office after that week's work, but had none to draw because of the custom of keeping a week's pay "in hand". This is necessary owing to the need for counting up the coal accounts and other bookings, but it is a severe handicap to a man who has been away from work for awhile or a stranger to the place who has no way of paying for his keep for the first week. One has to wait for a fortnight before getting the first pay; then there are two weeks' lodging and living to be paid. It is a safeguard that a man cannot leave his work without notice unless he is prepared to leave a week's pay behind. The sole advantage I found with this system was at strike time, when we had a full week's money to draw several days after we had finished work.

During the second week I worked by day. This was more pleasant, as I was home and washed by half-past four, and could go out in the evening content that work was over for that day and that I would have a night's sleep before I worked again.

The men in the headings were the only colliers that worked by night. The stall-men worked only by day. The repairers, timber-men, rippers, and labourers were working by night so that they should not delay the output of coal.

Coal was the only thing that mattered by day. A " turn " was kept so that each collier had his tram in rotation and with regard to the conditions under which he was working. A collier with a helper could claim a tram in every turn if he had the last one full. A collier who worked alone could only claim a tram on the odd turn—first—third—fifth. This turn lasted all the week, and each collier had to mark the number of trams he had full as well as his marking number. All the others could see this mark, so no " dodging the turn " was allowed.

I have met people who imagine it is possible to do any amount of work-dodging in the darkness of the mine. Some may be done, but the work is so measured, allotted, and weighed that a man must show some work before he can claim any pay. The men are good guardians against scheming, also, for what one man dodges another must do, often in addition to his own. Men who do not do their share are treated with contempt and are given nicknames, such as " Shonny one tram ", by their fellow-workmen, who are usually too ready to pour out their sweat and their blood.

During the last five years I have known three men who have died after no apparent signs of illness, and the doctor has stated that they died because their work drew all the moisture from their bodies in the form of sweat and it was not replaced.

Thursday night is "jib night" at the collieries. This term is supposed to refer to the men's faces when they get the docket which shows how much pay they should draw on the morrow. Some of the men looked at the paper for a few seconds, then pushed it into their pocket and walked away. Either they had received somewhere about what they expected or they were too disheartened to make any fuss. Others would jerk upright and stare at the paper with their mouths open and their eyes bulging before they made the usual comment of:—

"Well, I'll be damned! Just look what the blasted old sod have put in for me!"

I had expected to find men with a prayer on their lips when at work underground, but I soon discovered that there was a lot of swearing and very little praying there. Some of the men were convinced that the sledge would not hit so hard or the horse pull so well if a little swearing were not done, just to help. "Erny" was by no means an isolated example. When he hit his head against the top he swore, when he pinched his fingers with a stone, or hit his shoulder against the side, or knocked his lamp in the dark he swore, and as one or other of these things was happening to him all his working day, it became second nature for him to swear at anybody or anything. One night he swore in the railway waiting-room and was summoned. He vowed it was unjust—"because I wasn't swearing proper, like".

"What did you say, then, Erny?" I asked him.

"The train was late, see, mun?" he explained, "and I only asked the porter, 'When the h—— is that b—— train coming to this b—— station?'

Just like a chap would ask the b—— haulier about the b—— tram, like. An' I got to pay two b—— quid for that. B—— shame, I calls it."

Yet I do not believe he would willingly have hurt a fly, and I have known him give his last sixpence to a tramp.

On Friday of that week I saw the "measurers" for the first time and learned a little about the method of paying for colliery work. In small seams such as this one it was necessary to rip the top down, so that height would be made to allow of the passage of horses and trams. This ripping was paid for by the inch of thickness per yard forward. About six feet wide would have to be ripped. Twopence an inch was the price here, so that if we ripped one yard forward and it was two feet thick, we got four shillings. In this price was included the time for boring holes, the cost of powder, the use of your own machine, and the time taken for breaking the stones small enough to get them into the waste behind and making a wall of them so that they would support the roof. The posts that support the sides of the roadway—none others—are paid for at sixpence each. Then there would be the price for filling the larger coal. Any coal that falls through the smaller screens is not paid for. I believe the price at this colliery was two and fivepence a ton. This price varies at almost every colliery according to the thickness of the seam or the difficulty in working it. The big seams of nine to eighteen feet are lower in price than the small seams that range from a yard down to eighteen inches or even a foot in thickness, or thinness.

Each colliery has its own price-list which is supposed to govern payments. The price on these lists is constant, but after the items have been totalled up there is a percentage added, which varies according to the agreement in operation at that time. Thus, when the percentage rate ruling the coalfield is twenty-five per cent., five shillings would be added to each pound earned on the price list.

On that Friday morning I had noticed that my mate was continually adding little items to a list he had chalked on our curling-box. These curling-boxes are like very wide and large sugar-scoops, and are used for carrying the smaller lumps of coal. John had written several queer figures and words on this tin box, and seemed quite satisfied with his markings, then he went on cutting coal.

Then some lights showed higher up the heading, and John said, " Here they comes. The measurers," in a tone that gave no hint of welcome.

When the new arrivals got near to us they made a deal of noise. One shouted, " Where's your ripping mark, John? " at the same time as another bawled, " How many posts here, John? " There were five of them, and all seemed anxious to talk at once. But John was not perturbed. He was wise to that game, and rose from his knees slowly while he waved his hand to the side and stated, " Six posts."

Three of the newcomers started to speak in English and the other two in Welsh. John pushed past them, and did not say a word in reply. Then he knelt down before his chalked list and called out each item singly. He refused to mention the next until he was sure the last one had been booked. One man measured the rippings and shouted his findings to the under-manager. That did not suit John. He insisted on another measuring, and held his lamp to the tape so that he proved himself to be right. During this argument the fireman who was scratching the posts to show they had been paid for had been shouting that he could only find five: " What d'you mean by this game, eh? There's only five new posts here." John walked across and counted the six to him very deliberately.

Another fireman was bothering John about the football team, while one of the others tried to get him to argue about the Cymanfa Canu—or singing festival—that had been held on the previous Saturday. I do not know what the other said in Welsh, but he talked with scarcely a pause. They were wasting their breath, for John closed his mind to them until they had finished his items.

When they saw John had finished his list, they paused, and my mate—seeing the under-manager's book still open—suddenly remembered another thing.

" Aye," he said, " the coal's sticking to the top pretty bad here, aye it is. There's allowance on the list for that, ain't there? What about some for me this week? "

The under-manager frowned, then asked sharply:—

" How many trams have you got full? "

John gave the number—we had less than the previous week.

" You're down," said the under-manager: " we daren't give any allowances unless you've got enough coal to cover it."

" And last week," John answered very slowly, " you told me as you couldn't give me nothing for it because I had a lot of coal full."

The under-manager snapped the book closed, rose

from his sitting, and walked away, with the firemen hurrying behind him. About one minute later we could hear another argument starting in the next working-place.

John told me afterwards about their methods. He was a Welshman, and had to choose his English words carefully. He considered them bullies who were intent on avoiding payment for work done. Their intention was, so he explained, to badger the workman in every way, to talk about his interests, to frighten or fluster him—anything to make him forget some item in the measuring. After they had once passed it would be very difficult to get paid for it. It would be claimed that the work had been measured and that ample witnesses were at hand to show the man had had a chance to book it, but had not done so.

I noticed on that first Friday—as I have done ever since—that there are plenty who solicit help from the miner. Beggars were outside the pay-office, holding out laces, moth balls, or lavender packets—anything for an excuse—and almost every pay-day some organisation had an appeal or flag day.

There was a large group waiting to see the manager to get their money made up to the minimum wage. They had to go into his office and answer questions as to why they had not filled more coal, and so on. There was nothing courteous about the questions or the way they were put, but the answers had to be given very humbly. Even humility did not always succeed, for I know that many men had to go home short of the wage that the law is supposed to enforce.

The method was to promise to give it next week, then the week after, and on in that way until they would argue that it was so long ago that they could not recall the exact happenings. Some men went without what was due to them because they feared victimisation. Another man who was afraid to ask for what were his dues kept count of the amount he was short, and seemed to get some satisfaction by telling us, confidentially, that the company already owed him about forty pounds.

I do believe that they tried to avoid letting any man go short of his pay if they judged him likely to make a fuss over it, but occasionally a man would refuse to let them avoid payment, and the committee would be compelled to take action. This would take time, and after a series of delays and evasions the case might be taken to court. The usual method then was to pay on this case before it was judged and leave the others still unsettled.

The man who forced a payment in this manner was wise if he tried to find a new job, and it would be very probable that he would have to go some distance, for he would be marked in that area.

There are so many ways to rid themselves of an awkward man. It is a long distance in to some of the workings, and stones or slags are often piled on the side of the roadway. Just a few stones pushed into a tram of coal in the darkness mean a case of filling "dirty" coal when that tram comes to daylight. Dirty coal means lost orders. How can a committee defend that case? How can the man prove he did not fill it?

Or a collier may be kept so short of rails that he has to throw his coal four or five yards farther than the man in the next place. He may be kept short of trams or timber, or put in a place where the coal is very stiff or there is a lot of water. The man does not—he cannot—fill so much coal as those in the other places, therefore he is lazy, is not trying; and with his job goes his house, very often. The tied-cottage system of the farms has its ugly sister—the house owned by the colliery company.

My mate had made a little more than the minimum wage that week. He had the main deep heading, and it was at a good price. He had one little fad of his own, however. He did not like paying me unless he included a drink—just one half-pint, for he was a temperate man—in the business. It was awkward in my case, for the smell of that drink would have meant disaster in my lodgings. I had to have my pay, and John was determined. I had the drink, and another half-crown as extra pocket money, so I bought some scented sweets as well.

I was getting quite fond of John by now. He was such a sincere type and tried to explain things to me and make me do them myself.

I must confess, too, that I had not heeded Tiger's warning about " them Welsh gels ". I was finding one dark-haired and vivacious young lady very attractive.

There may have been a deal of truth in the explanation of an old Herefordshire woman concerning why her sons had married Welsh girls:—

" Diff'rent sort of gels they'm be up there," she stated. " Thes'n round of these parts do say as thay dunna want thay chaps when thay do, and the Welsh 'uns do say as thay canna live a'thout 'em. An' the chaps do just naturally believe 'em; so thay do."

Anyway, I was settling down. Wales was becoming quite attractive, in a way that Jack did not welcome.

CHAPTER FIVE

I SAW very little in the way of accidents for the first three months. This was a small colliery, working a seam that was considered safe.

I remember having to work a hand-pump by myself one night. A serious accident had happened within a few yards of that place during the day before. I was on the jump all night. There was a wide crack above me which went up and up until the light from my lamp would show no farther. I was too inexperienced to know whether this roof would stay up or be likely to fall. I could smell the used iodine and see some pieces of bandage. Morning was very welcome, once again.

I had been working there about three months when a man named David Jones came to work behind us. He was skimming the edge of the coal off the side of our heading until he had enough width to turn his own road.

One morning he was kneeling down holing when a stone fell slantwise from the roof. It was not a big stone—no more than twenty-eight pounds in weight—but it was sharp-edged, as were most of the stones in that seam.

David was a loud talker, so we clearly heard his opinion of things.

"This blasted hole!" he complained—"the damned stones here do fall like lead. They do hit a chap so hard he could think his foot was off—aye, indeed."

Dai turned to look at his right foot, then gave a loud shout of:—

"I go to hell. It is off, too."

I remember them placing the foot alongside Dai on the stretcher, and his complaint that they had bought that new pair of shoes only the week before.

So, because of his wooden leg, Dai Jones became Dai Peg, and some months later it was necessary for him to ask for another job at the colliery. Dai went to see the manager on one of the few mornings when the manager was in a good humour, and in any case Dai had a claim on that firm for light employment—he had been crippled for them.

Dai looked through the little window, and could see the manager sitting at ease in the office and smiling—yes, actually smiling—back at him. Dai judged this to be a good omen and inquired:—

"How go, Mr. Thomas? How's it looking for that b—— job as I'm supposed to be having?"

The manager was in an extraordinary humour that morning. He got up and walked across to the window.

"Now, Davy; now, Davy," he remonstrated, "that won't do, you know. That's not the way to ask for work. You must ask in the proper way."

"All very well for you to talk," argued Dai, who never twisted his thoughts or words, "but if you was standing out here in my place you'd feel diff'rent, you would."

"No, I wouldn't," the manager objected: "it always pays to do a thing the right way, and it costs nothing."

"What do you call the right way, then?" Dai asked.

"You come in here for a minute and I'll take your place and show you."

Dai went inside and sat down grandly in the easy-chair, laying his injured leg on the low table with the wooden part pointing like a rifle towards the door. He enjoyed the position, and took a cigarette from the box on the table to complete the illusion.

"Good morning, Mr. Thomas," sounded the manager's voice, so humbly, from outside the window.

"Mornin'," Dai growled between two deep puffs at the cigarette.

"I called to see if you have got a job of work that I can manage to do, please, Mr. Thomas?" The manager made this appeal sound as beseeching as he could.

Dai did not move from his comfortable position. He made the tone of the words convey the definite nature of his reply.

"No," he answered, "there's nothing here for you. There's too many idling about here as it is. Clear off."

As this was an almost exact repetition of the manager's usual reply to a request for work, it is possible that he may have altered his ways after that, but I never heard that it had that effect.

I found that there were many pleasant walks through the valley. There was an old coach road almost overgrown by the trees, and higher up on the mountain a Roman road that travelled along its breast. How I used to enjoy that climb upwards, with the keen wind smacking my cheeks and nipping off the tops of the dry grass and blowing them into the cracks of the stone walls that were the boundary-marks on the mountain. Every walk was an expedition in which a mound of grass might be a Roman grave, or a pile of stones a ruined fort. I was not so lucky as some, for I failed to find any tokens of the Romans, although I did search. There was a ruined Abbey at a distance of about six miles. I stayed there alone late in the evening to watch the moonlight shadowing the pillars and whitening the ground amongst the fallen stones like a silver sea.

There were waterfalls as well. One in particular, where a mountain stream rushed out and fell for eighty feet. It was a brown torrent in winter, roaring in fright when it took that terrific leap and carried trees and sometimes sheep with it on to the rocks below, but in the summer it was a wide veil of water, the spray from which kissed the primrose roots amongst the rocky sides to new life and beauty. This fall was not more than a mile from the drab village, but I took several natives to see it for the first time.

It was natural that I should be interested in the horses underground and their treatment. Horses were a part of our lives in the country. I had seen them foaled and reared. I had helped to break them to harness and to work. At the age of sixteen I could handle three young horses at any of the work on a farm that horses do.

I have read statements in which it is argued that there is no cruelty to horses underground, that they are better off than those that work on the top. These writers may believe what they write, but I do not. Probably they had their information in the colliery stables, and there is very little cruelty there, unless one counts as cruelty the absence of daylight, and fresh air, and green grass.

It is in the actual working that cruelty occurs. I know that strangers never see this happening, but I have seen much of it during the last twenty years.

I have seen a horse that was blind in one eye rubbing its nose against the sharp stones so that it could feel the way to turn, and being whipped for not coming round fast enough. Every day horses are worked for sixteen and more hours straight off. What I have seen I know to be true—and it happens regularly. Very often they have no water at all during those long hours in the hot places of the mine. Within the last years I have seen horses drop from exhaustion several times during a shift. They were stripped of their heavy shafts and curved tram " gun " and whipped until they staggered up to have the harness replaced, and the rush to get coal back was resumed.

I do not believe that the average colliery haulier is so good a horseman as is the farm waggoner, nor anywhere near so good. They have not been so well trained to the needs of their charges.

All the blame is not on the men, for they work under a bad system. The conditions make humane handling difficult. Often a haulier is compelled to take a horse out again when it has only just been brought in from working a shift. No allowance is made for the fact that he is working a tired horse. He must return with his trams as quickly as the man who has a fresh horse. Often a man must choose between forcing an exhausted horse and being sent home and on the dole—with consequent suffering to his family. He must choose whether the horse shall suffer, or his wife and children.

Probably he would even be penalised for a period, and would not have the dole because he had been dismissed; with a pretty sure charge of being lazy and being insolent to the officials.

Its costs money to rip the top and the sides underground, and so no more height or width is cut than is absolutely necessary. When the shift begins there may be room for a horse to travel along the roadway and some inches of height clear above his head. Sudden " squeezes " occur, during which the upper roof moves and the weight comes down so heavily on the roadway that the sides cave in, the bottom is forced up, and the roof is pressed down. After one of these squeezes the total space left may be some inches less than the height of the horse, yet he must be driven underneath it. The horse comes to this low part, and when he goes under it is bound to bend, almost until his knees touch the floor; but they do not often have enough sense to bend until the skin has been sliced from their backs by the jagged stones of the roof. Again and again the horse is forced past this low place, each journey increasing the number of gashes on his back.

After the first journey the horse makes a desperate rush to get through quickly and lessen the time of agony. The haulier, perhaps taken by surprise, is nearly jerked under the tram by this unexpected rush. He concludes, " This blasted horse is gone rocky. But I'll steady him, though. Just watch if I don't." He puts the chain of the leading-rein through the horse's mouth and holds him to a steady pace, especially under the low place, and so prolongs the agony, although he may not realise it.

Then the horse, with stones ripping into his back and his mouth twisted by the chain that stops him rushing away from this torture, does the thing that his instinct tells him is the only way to get free—and kicks. After that first kick he is marked as a dangerous horse, a kicker and rusher, and is treated as such. This is often not such a bad result, because a vicious horse is avoided and not worked as regularly as a quiet one. A vicious horse is doubly dangerous underground, because there is little room to get away from one. I know that a man can refuse to work a horse under a low place, and is entitled to take him back to the stable, but he had better not be dependent on colliery work for a living if he does things like that.

I remember one horse which was quite docile outside, but when taken into the colliery he became a demon. No two men could handle him, and at last he was taken out to be sold for use in a coal-cart, where he became quiet at once. A most sensible horse.

The usual comment of the miners when they see a fresh horse taken underground is, "Poor devil! Another that won't be long before he's meat for the dogs."

I understand there are inspectors of pit ponies. I have never yet seen one. I was talking to two friends a few days ago, and I mentioned this to them. Each of them had worked in the mines more than twenty years, but neither had seen a pony inspector. As for the mines inspectors, one of these friends had seen one on two occasions, the other three times; while because I had been concerned in several fatal accidents during that period, I had seen a mines inspector eight times.

Many of the men hated to witness the cruelty to the horses, and I have seen more than one fight over it. These men could not alter the method that forced tired horses to work on, but they did interfere when the brutal ones beat their charges with thick wooden sprags or kicked them with heavy nailed boots. One of the most assiduous protectors of the horses was named Jack Edwards. It was through his habit of bringing apples and little tit-bits from the garden for them, and his determination that they should not be beaten, that I first took notice of this Jack. I am sure he would have fought the battle of anyone whom he thought weak and needy.

I had been working there about twelve months when I spoke to him first. He was some years older than I was. I think he must have been about twenty-six. I do know that he had been courting for some time, and would have been married, only things were difficult at home. He was good to look at, well built, had clear skin with a tinge in his cheeks, and dark hair curling about his head. Jack was clean in every way.

Jack and his father worked together—they were repairers. The father had some chest affection that made breathing difficult at times, and often he could only manage to stand up and watch his son at work. But they were always expected to do the same work as two strong men, so Jack had to do most of his father's work as well as his own. He did it cheerfully. There was an invalid sister at home, as well as a brother who was costing them a deal of money to keep at his studies. So Jack did double work and even went short of pocket money. One night he was making a place to set up timber on one side of a tram while his father was struggling to do a bit to help on the other side. Suddenly the father heard a strange sound and commented:—

" D'you hear that, Jack? What was it? Sounded as if something was a'sliding, like."

He heard no reply from his son, and so he straightened up to look over the tram to where the latter was standing. As the light was so feeble, he could only just distinguish Jack's features, but his eyes were looking towards his father, and it seemed they were trying to convey the message that his mouth could not speak. While the father stared at him, Jack seemed to shiver all over, then slump forward. The father stumbled round to that side of the tram, and found that a large stone had slid from the side and its sharp edge had caught Jack against the tram, almost severing the upper part of his body from the lower.

I dreaded passing that spot for weeks afterwards. I had to go that way, but rushed past as swiftly as I could. That is one effect of an accident that I had not considered before that time. Someone must get the stone off a dead or injured man; someone must break that stone and clear it away; someone must put the working-place in shape again, and someone must work there every day and every hour afterwards under the same piece of roof that betrayed his friend or relative, and with the stone in sight and mind all the time.

What shocked me most about this accident was the method of paying compensation. When a single man with no dependants is killed, the only payment allowed is for burial expenses. Jack's father was working, so he was assumed to be the breadwinner, although all the men knew that Jack had carried his load as well as his own. The father never did work there again. He could not have done one hard day's work on his own, and probably the shock he had endured would have finished a much stronger man.

Because he was classed as single, Jack Edwards of the noble ideals and kind actions was counted to be worth eighteen pounds.

On the night of his death I heard a fireman warn a haulier who was noted for his brutality to be careful how he handled a new horse. He did not ask him to stop beating this horse; he warned him to be careful not to let an accident happen to him, because he had cost the company forty pounds. I did not relish the idea that I was worth only half as much as a horse.

I had been given two rises in my pay, and was getting thirty-six shillings a week, apart from a few odd shillings as extra pocket money. I was of a type that spent very little on enjoyment, and had been saving steadily. I was nineteen—a full man in my own opinion—and although I would not be old enough or experienced enough to claim the wage and working-place of a fully competent collier until I was twenty-one, I was all eager to sample the independence and joys that should come with manhood.

My lodging was excellent, and so was the home of my sweetheart. Yet we were as young lovers always have been: we longed for a home of our very own. The trees alongside the woodland walks heard our whispers of the furniture we would buy, and the grey stones in the houses were surely amused at our whispers of quiet rooms and places where no one should intrude.

Besides, when a man married he sent his own value up to about four hundred pounds. He was worth something. We allowed nothing to interfere with our planning, and so the date was fixed.

Jack was getting very little of my company, so he was lonely again. He did not make friends easily, and the people he lodged with were shortly going back to live in the Rhondda Valley. Then Jack had a nasty gash down the side of his face during the last shift before we went home for one of the Bank Holidays. When his relatives saw his injury they begged him not to go back to the mines again.

They did not have to coax a great deal. Jack stayed in Herefordshire, and is still there.

I returned alone to Treclewyd, and spent the next three months in saving hard and making preparations for getting married.

CHAPTER SIX

When the service was over and the folks who held the gates had been paid to reopen them; when the children who held the ropes across the road had dropped them so that they could scramble for the coppers we had thrown; when I had heard my wife addressed as Mrs. Coombes for the first time and thought how strange it sounded; when I had eaten about five mouthfuls of the wedding breakfast and we had driven through the sunshine of that September morning to the next station—we avoided Treclewyd because a crowd was waiting—we sat back in the railway-carriage and considered what had happened.

The old minister in Treclewyd always concluded the marriage ceremony by saying:—

"And now, bachgen—and you, merchi—you have been made one in the sight of our God. So now you will both try to build a little nest for to live in and be very happy. You, bachgen, will go to work hard and bring all your money home to your wife as she is now; and you, merchi, will try to make the house nice for your husband when he comes home and will try to do the best possible with his pay and give him just a little bit back for pocket-money. Then everything will come all right. Yes, indeed, it will. So, good luck to you both."

His advice was in line with our own plans. We had both been taught to work hard, and neither wanted luxury or had expensive tastes. We faced the future quite prepared to sacrifice for the little home we had talked about so much.

We had a short honeymoon amongst the loaded apple orchards of Herefordshire. While we were in the train on the way back to Treclewyd the explosion at Sengenydd happened. The rest of the journey was saddened by this, and everyone was gloomy at Treclewyd when we returned. The disaster was so near, and everyone realised that it might have happened to them and that Treclewyd might be a village of lowered blinds and slow, black processions.

My wife was a native of the village, and Welsh was her natural language, but there was nothing lacking in the speed and fluency of her English. She had a large number of relatives in the village.

We had been promised two rooms in apartment in one of the streets, but found that we would be compelled to wait some weeks before these rooms were vacant. It had been arranged that we spend this waiting period at the home of my wife's parents.

It is the custom in these parts for the young couple to live this way for a while, and use the rent they save to help in the buying of furniture. It is very often the only help they can expect.

My father-in-law was in a privileged position. He had been elected workmen's check-weigher some years earlier. He was the only man on the colliery where he worked who did not receive his pay from the colliery company. His wage was paid by means of a small deduction from the earnings of each collier.

As check-weigher he was supposed to be independent of the company and to watch that the man they employed to weigh the coal allotted the miners their fair weight. It was necessary for the check-weigher to be an honest and keen man, and my father-in-law had held the respect of the men for many years. He was also Secretary for that Lodge of the Miners' Federation, and handled the men's side of compensation cases. Every night he spent in the front room busily engaged with complaints or claims from the workmen until he went to bed.

I was eager to make a good start in my new life, so I was up and away to work early on the first morning after our return. When I arrived at the mouth of the level the under-manager stopped me and told me our old place had not been working since I went away and would not be working for some months. He wanted me to go to work in another district. My usual mate was ill and would not work for awhile.

I had heard a lot of talk about this other district and the conditions of work there. I had seen something of it one night. The coal there was very thin—under two feet in thickness—so that it was necessary to lie down flat when working. It was very wet—so wet that there seemed to be water dropping from every inch of the roof and the water was flowing to a depth of about two inches over the bottom that the men had to lie on. It meant that they had to lie in water all through the shift and be wet to the skin within a few minutes from starting.

I realised that the under-manager thought it was a good chance to force me to go there to work. He guessed that I would not wish to have any trouble when I was only newly married.

I argued with him that as others would not go there to work, I did not see why I should. He replied that it was the only place he had at present, but if I would try it for a bit, until something else came—I had heard him give this sort of promise before—he would see what he could arrange.

" But it means that I'll be soaking wet all day," I protested.

" There's allowance for that," he said.

" How much? " I asked.

" Well, it's threepence a day extra." He was a little ashamed to state the value he set on a soaking back.

I had made up my mind, and told him so. I was not going: it was too wet.

" Not making a good start for married life, are you? " he sneered.

" Might do worse," I replied; " it'll be better to do that than be in bed with rheumatic for months, the same as some of the others you have put there."

" Make your mind up about it," he said, " and let me know when I come out from seeing the fireman. I won't be a minute."

He went into the cabin, and while he was there I considered three things. I had my old insurance card out, and my new one was not handed in. I had no pay in the office, because I had not worked the last week. More important than all, I had a chance of work at another colliery. I had made friends with another young fellow a little older than myself. He had been given a stall at another colliery, and wanted me to go on shares with him. It meant more money for me, and only the previous night he had been coaxing me to come with him.

I felt it was a good chance, and I wanted to show that under-manager that I was independent of him. I turned sharply, then, kicking up as much of the plentiful dust as I could, I went homewards. The check-weigher came to the door of his cabin and asked if there was any trouble.

"None at all," I assured him—"only the under-manager made a mistake."

I emptied out my tea and strolled slowly down the incline. Treclewyd was not properly awake, and few people were to be seen. I had indeed made a bad start in my married life. My wife was busy at her part. She had made her mother stay in bed, and had scrubbed the kitchen and the passage clean. She was kneeling on the flagstone outside the front door when she heard slow and heavy footsteps and, looking up, saw her wage-earner returning self-consciously down the street—without a job.

I felt better after I had explained things and my behaviour had been approved. I felt better still when I saw the pleasure with which my friend received my news. We went to see the officials at his colliery that evening, and I was given permission to start. I begged, borrowed, repaired, and bought enough tools for the beginning.

I should have to rely on the extra experience of my new mate in work, and I was not old enough to claim a stall of my own, but they did not seem to worry much about those things at this new colliery. There was no fuss here about signing on. I was told to call in when it suited me and stick my name down in the book.

The name of my new mate was Tommy Davies, but as he had come from Swansea some years before, he was scarcely ever called anything but Tommy Swansea. Any woman could have been jealous of his complexion. His skin was clear and white, and his cheeks so red that they appeared to have been painted. I never knew him have any need to shave in all the years I knew him. He was over five foot ten in height, slingy in build, and very strong and active. Like myself, he rarely drank intoxicants, and never smoked, so that when the other workers paused for a brief whiff—this again was a naked-light colliery—we kept up the rush of coal-getting without the least slackening.

Tommy was a great worker. Hour after hour he would bang the heavy bottom mandril behind the slips of the coal. We learned to lift together automatically, to change our working positions to suit each other without saying a word, and to vary our jobs so that the change should give some rest. For some years we shared out troubles and joys, good weeks and bad, fears and ambitions—we were real " butties ".

I was downstairs before half-past four on the morning that I started to work at the new place. An early start was needed, because we had to climb nearly three miles of mountain before we got to the work.

At half-past five my new mate called. He was not late that first morning, but it was too much to expect that he would be early. That was Tommy's one bad habit: he would not get to work in time. He used to climb the slope at a tremendous speed, with his cap in his swinging hand, the steam rising from his thick hair, and his cheeks glowing like red jam-apples; but the hooter would always go just before he arrived at the level-mouth, and, with their rigid insistence on time, the officials had a job to avoid treating him as they would have any other man. I remember the usual dialogue.

" Now, Swansea; you're late again, you know. The hooter's gone this three minutes. This won't do; you'll have to go back." The under-manager would glare at Tommy in an attempt to overawe him.

Tommy's blue eyes would sparkle with health and mischief when he replied :—

" Sorry I'm late, David. And I did feel like filling a pile of coal to-day, too. We've got a couple of trams ready."

The official would turn his gaze on the mountain when he questioned :—

" Got a couple ready, eh? We could do with some extra coal out to-day, too."

" Well, there you are. I can't help to fill it if I go back home, can I? "

" No, I s'pose you can't. Clear off inside quick, then, and don't let any of the others see you coming in at this time. And for God's sake try to get here in time from now on."

Tommy very rarely did get there in time, however, and if he did start early he would dawdle on the way so that he had to make a grand rush at the end. One day the people where he lodged made an effort to cure him, and forced him to get up fully an hour earlier than usual. Tommy used that spare hour to eat a double-sized breakfast, and was so full that he arrived later than ever. They did not try that remedy again.

He had allowed bare time to get there on my first morning. He whistled outside our door, took a share of the tools under his arm, and warned me that we would have to travel or be late. We did hurry, with the load of tools under our arms, a hearty breakfast inside us, and the sharp slopes of the mountain to stick our heavy boots into.

Outside the village we struggled upwards through a lane that was almost a tunnel. It had a floor of solid rock and sides of dry stone walls about a yard thick and six feet high. We clattered up this lane for more than a mile, then crossed a brook by a long plank to reach the open mountain. It stretched as far as I could see. There was only one house—a farmhouse—to be seen at all, and but a few faint tracks to guide us. This sharp upward slope was covered with a coarse brownish grass that grew very long. In the hollows were bogs, so that it was dangerous for a stranger to cross alone. That was the reason most of the men carried sticks to help them probe for solid footing, as well as to ease the climb and prevent them slipping.

Nearly two miles up another stone wall was built across the mountain. This was the boundary of one of the sheep-farms. The main path went to a break in this wall, and there was a gate which could be opened if you lifted it high enough. Some of the stones had been pulled off the wall to make seats, and this spot was known as the Whiff. A long row of men—more than sixty in number—sat alongside this wall, resting themselves for the final climb. Old Woodbine packets and matchsticks littered the ground, and were mixed with the dirty pieces of wool that had been left about. No one seemed to be talking much, as they needed to recover their proper breathing. There was a blue immensity above, with the morning sun shining on the great mountain where the men seemed to be black dots.

We had scarcely sat down when a horse with a man on its back showed itself over the skyline. Another mounted man followed, then another, until about thirty could be seen one behind the other. Their appearance was the signal for the men at the wall to re-start the climb. We had plenty of help with the tools from this spot. We stumbled over the rough grass downhill for a short way, across another brook—on a fallen tree with a wire rope for a hand-rail—then through the yard of a tumbled-down farmhouse where the garden path could still be traced, across a narrow bog, and sharply upward until we stepped on to a double set of narrow-gauge rails on which at intervals were wooden trams fastened to a thick steel rope. The hauliers arrived there at about the same time. They passed at a slow trot, riding bareback, with only a bitless bridle and a collar on their horses.

The morning greetings were well spiced, and they rode on to where a huge black tip spoiled the continual brown-green of the mountain. I noticed that nearly all my companions were young men. All were in training from that walk; not one had any trace of stoutness.

I had not seen one tree or place of shelter since we left the lower part, and judged that it would be a devil of a place to reach in the winter. I found myself right in that estimate—later.

This was a new colliery, for it had not been working two years. The engine-houses were roughly made, and the cabins for the outside workers were simply broken trams pushed up on their ends. A great amount of rubbish had been brought out already, to judge by the size of the "muck tip". They tipped nearly two hundred tons of rubbish every night. The manager, under-manager, and the firemen were all under thirty. They were keen, but had a difficult seam to handle, and were handicapped by not being allowed much capital to work upon. They had the reputation of being willing to pay for work done if they were allowed to do so. That was the reason that men were willing to struggle across the open mountain in all weathers. In the summer it was risky to go without a coat, because five minutes of the driving rain up there would wet a man to the skin, while in the winter everyone took oilskins and leggings with them each day, yet went home wet very often.

Tommy and I went inside a stone cabin where some dirty books were placed on a desk made of two rough boards. Two wooden sleepers and an upturned bucket were all the seating accommodation. A big fire burned in the grate despite the warmth of the morning, and the ashes needed clearing away. Four cans of tea were keeping warm—very easily—on the hob. The only bright colours in there were the two painted stretchers and the splints that were standing on end in the corner.

The manager came in; he had to bend under the doorway. He was perspiring, but was quite friendly.

"This is the new mate you spoke about?" he asked Tommy. "If he turns out as hot as you are on filling coal, we'll be mighty glad to have him. We want chaps here pretty badly that can fill their whack."

We went from the sunshine into the wet darkness. It had the same kind of notched timber protecting the mouth as had the other level, and similar stale-smelling and oil-covered water flowing out between the rails; but as soon as we got under the mountain, things were strange to me.

The seam worked and the type of roof were very different. In this working we could see that distant circle of daylight almost until we reached our working-place; and very enticing that circle seemed when we looked back. This coal-seam was not a true seam: it was what is known as a rider to the proper seam, and we were not more than sixty yards below the summit of the mountain. This was no advantage to safety, for the roof was not so solid as if we were working the lower seams. We were able to tell when it was raining outside because the water ran down to us through the cracks in the top and the roof started to melt.

My eyes had not become accustomed to the change from daylight to darkness before we reached the working-place. Tommy had warned me that the ventilation was very poor in this colliery, but said they hoped it would improve before long.

As soon as we got to our working-place I found it hard to breathe. I was gasping for air, and my lungs felt tight. Almost all the miners used candles. These were stuck in a hole in a sharp spike and the spike was forced into a post. If we moved the candle the least bit it lost its light. It would go out without flickering at all; it would lose its flame slowly until the end of the wick was but a sort of red fuse; then even that would fade. It was useless to try to light a match to restart it burning, for the match would not flame; the top of it would smoulder, then turn black.

We had to grope our way back to the main level, and when we got there and the ventilation was better we relit the candles and carried them very carefully, slow step after slow step, back to the coal-face again.

Hour after hour the candle-flame stayed still in the darkness—it never moved the least bit; it never flickered. One candle would last for twenty hours. Men were trying to work where there was not enough air to keep the candle burning.

Near our tram we had to fasten the candle upside down, so that the drops of grease should fall on the flame and so keep it burning. At the lower end of our twelve yards of coal-face the candle would not keep alight, no matter what we did, so we had to try to do our work helped by a candle that was not giving one quarter of its proper light and had to be left yards away.

We hit our hands and shins, and very often our heads, against coal and roof that we could not place until we hit against it.

I had worked nearly an hour when I had a feeling as if my head was bursting. My eyes were burning and my lungs seemed to be lifting lead. All at once I was sick; I vomited without knowing I was doing so, and for the first time in my life I had a headache. It was a terrible ache, too—like some-one banging on my forehead and the balls of my eyes. Tommy noticed me swaying, and stumbled across to me.

"I've had a guts-full of it myself," he admitted; "let's go out and get some fresh air, before we leaves it too late."

When we started to walk, I found that I could not balance myself, so that walk became a tumble, during which we steadied ourselves by leaning against one another. We could not balance on the rails to avoid the water on the main roadway. We walked through that water up to our knees; indifferent to all else except the urgency of getting out to daylight.

When we came to a piece of low roof, we could not bend under it: we just walked against it and tottered forward when we had been knocked down low enough.

Nearer the level-mouth we could see the figures of other men going out, and we met two going back inside. They were refreshed after a spell in the fresh air, and told us wistfully that "it was grand to be outside".

The officials understood that men must come outside often during each shift—they had to, or die, and it had nearly happened that way more than once—and they only watched that no man stayed outside too long.

How lovely it was to sink down on the hard stones of the tip and let the sunshine kiss our bodies and the sharp mountain air invade our lungs!

Nearly every day some of the men collapsed, and I frequently saw artificial respiration being tried. One day they failed to get one man to recover properly, and they had to carry him down the mountain as fast as possible. Before they reached the village he regained consciousness and wanted to know why they were carrying him. He insisted on walking the rest of the way, and was in work next day.

Until I went there to work I did not know what a real headache meant; but I found out, for after that I had one during every hour of every day. As soon as I bathed I would drop on to the couch, and there I would stay until my wife shook me out of my stupor so that I could have supper. Next morning came with the ache still in my head, and when we got back again in work we found that the smoke from the blasting-shots we had fired the night before was still in our working-place. After sixteen hours there had not been enough fresh air in those workings to move the powder-smoke away.

These conditions lasted for week after week. The officials did include some allowance in our pay, because it was impossible for any man to earn his money—I mean to do enough work to make his money clear without assistance. I was then receiving the full minimum rate of seven shillings and fourpence a day.

Not long afterwards they found a solution to the

lighting problem: they began to use a small carbide lamp, an American patent, which could be held on the cap and would burn with very little air. It was possible to do more work with these, but we had to buy them ourselves, and they cost five and ninepence each. We had to get one each very quickly, because it was impossible to move with the candles without losing light, and these new lamps were so bright that the men who did not have them felt the uselessness of the candles more than before. Even these lamps could not give out their usual white light: the lack of air made them have a bluish flame.

One of the men managed without one of these lamps for weeks, until the others complained that he was always messing about in their way so that he could use a bit of their light. I think he was afraid of his wife, because he wanted a lamp badly, yet had to manage without one. He was the last to use candles. He was often teased about whether "he was looking for fleas with that night-light?"

He did arrive at last with a new lamp. He was proud of this lamp, but said that his wife had kicked up a shindy about the cost. The first time he put that lamp out of his hand a stone fell on it and flattened it. He was in a fix as to what he should tell his wife.

After rejecting a lot of useless suggestions, he decided to take the flattened lamp home and leave it on the doorstep, as it would be dark when he arrived at his house. That was the night for the weekly call of an insurance agent whom this man detested. He intended to talk about this lamp, and then go to get it for show. It would not be in his jacket, of course, but he had plenty of holes in that jacket that he could point to as deserving of blame.

Then he meant to find the damaged lamp on the doorstep and blame the agent for stepping on it. It worked well, for he dodged the blame, got his clothes mended, and his wife finished with the agent for ever. He thought it an excellent make of lamp ever after.

There were several collieries near to the village. One, at which I never worked, was the place which employed my father-in-law. This company worked another level besides the one at which my father-in-law was check-weigher. At this other level there were, of course, another weigher and men's check-weigher. He was their representative—chosen by ballot, and liable to be dismissed by them after a proper notice.

The men had been dissatisfied with the behaviour of their representative for some months, and at last they had a meeting to consider it, and decided, by a large majority, to do without his services. They gave him notice, and before it ended they elected another man to represent them as check-weigher.

The workmen relied on the law that states they are entitled to have one independent man to watch their coal being weighed, and they were paying this man to watch their interests. This question of correct weight of coal is important to the colliers, because the greater part of their wages is paid for the coal they cut, and if a hundredweight less than the proper weight was recorded for each tram, it would come to a considerable total where seven or eight hundred trams are weighed every day.

When some of the men's committee escorted the new check-weigher to his office on the Monday morning he was to start his duties, they were told that the company would not allow his appointment; that the new man was not to enter the weigh-bridge; and that the owners insisted that the old check-weigher must continue to act as representative for the men. The check-weigher may have other duties in addition to the coal-weighing. He can watch the men's side of a complaint about slag in coal, and usually takes a record of accidents, so that he can watch the compensation claim.

When the committee had recovered from the surprise, they pointed out that they were quite within their legal right, and argued that this fresh action was a complete breach of agreement. It had no effect: the company refused to alter its attitude.

The committee returned to report to the men, and they were naturally not disposed to allow the company to dictate to them who should see that they got fair weight. When all hope faded of getting their man in the weigh-bridge, the workmen returned to their homes.

The men from the other level—where my father-in-law was check-weigher—heard of what had happened through a message that was sent underground during the day, and they were very quickly outside themselves. The dispute did not concern their coal, but it did concern men who worked under the same firm, and a vital issue was at stake.

There was no work next day, or that week, or the next week. So the strike went on for three months, yet the men were only trying to get a right that appeared to be guaranteed them by law.

There were some men among the strikers who, by reason of having good places or several sons working, had been able to save a deal of money. I shall not soon forget how some of these behaved. They would call to see my father-in-law to find out the latest developments or have a little chat—any excuse. After they had gone money would be found in places where only they could have placed it. When we sent it after them they would deny all knowledge of it. It would be used to help in cases where the need was urgent.

One night when I returned from work there was excitement in the house. Five miners' agents were coming to tea after a meeting on the cricket-field. My wife was acting as hostess because her mother was too nervous to do so. So in that small house—one of ten with slate roofs and whitewashed fronts—I heard five of our agents, three of whom became members of Parliament, pledge themselves never to rest until they had secured justice for my father-in-law and the men who were with him. Vernon Hartshorne was among them and James Winstone. I thought that colliery company was in for a rough time, but the dispute still went on.

Meanwhile, there was no improvement in the ventilation at our colliery. I was feeling the effects of working there myself, and there was a ghostly look about many of the men who worked with us. Some of them lost weeks of work at a stretch through chest troubles.

We were working a sort of house-coal there. It bound closely together when on the fire, and even the small burnt well—it had so much tarry matter in it. There was a twenty-inch layer of top coal and a fourteen-inch layer of bottom coal. Above the top coal was a two-foot-thick layer of clod. This had to be worked away before we could get at the coal. This clod was dangerous, because it often fell with no warning and rolled a long way back. It had to be stowed in the " gob ", and we had to clear more of this clod than we got coal. In some places every scrap of the coal had to be wedged or blown loose. Where the coal could be worked with a mandril, the place was counted to be a good one.

There was a good price on the coal—three and fourpence a ton. This was only a temporary price; until the price list could be agreed upon.

Tommy was ill one day, so I was working by myself. I was pounding away at the coal with a mandril when I felt the piece in front of me suddenly slacken. I steadied myself and drove the mandril full behind the slip, for I judged that the piece of coal was coming free, and I wanted to get it to finish the tram as soon as possible.

The mandril was driven forward with all my weight behind it, and the sharp point scrunched into the coal. Instead of being checked by a solid wall, however, it went on easily, and I fell forward with what had seemed solid coal falling with me. I crashed forward and downward for some distance, then knocked my head against something with such force that lovely stars came out and danced around me. The blow that had provided these stars had blocked the burner of my lamp, so I was in complete darkness, and was puzzled what had happened, and very scared as to what would happen next. The darkness of the mine is indescribable. You push out your hand to feel it, and your hand goes through solid blackness. There is no gleam of light or even relief of shadow. The darkness weighs on a man; it is as much a torture to the eyes as blinding light would be.

I was completely puzzled about what had caused my fall, and knew of no rail or gob-wall to guide me back. I had no way of relighting my lamp, but had the sense to realise that I must not move away from that spot, or else I might never be found. I felt with my hand, and touched what I knew were stones walled on top of each other. I sensed a musty, stale smell that was quite unlike the warm, smoke-laden atmosphere of the usual workings. No one was working near enough for me to hear the sound of their mandrils. I sat still and waited. It seemed that hours passed, and that I must have been left for the day and the rest of the men had gone home. In reality I could not have been there more than half an hour.

Then I heard a voice calling. I answered quickly—and very loudly. It was one of the hauliers, and he was near enough for me to hear his complaint of, " Ain't you full with that tram yet, eh? What the heck have you been at? "

" Tram be hanged," I answered. " Come and see where I have got to."

" Where are you, then? " he asked, and I could see his light swinging about more than five feet above my head. I called again, and heard him swearing in his surprise and wonder. I had to call a third time before he located me and came to the edge to look down.

" Well, I go to hell! " He was emphatic. " What've you done now? And what are you doing down there, eh? "

I had knocked into some old working of which no plans existed. The old miners had been cutting the bottom under the coal instead of ripping the top over it, as we did. They worked to where the coal-seam jumped up slightly, then they must have stopped working suddenly, for a loaded tram was rusting into the rails near where I had dropped. They had cut bottom right up to the piece of coal which I had been working, so that I had fallen into a hole cut by a miner who had worked there last eighty years before.

We noticed another thing too. A current of fresh air was coming through these workings. The powder-smoke was cleared in an instant, and our breathing was easier. The haulier threw a handful of dust in the air, and it blew away, instead of hanging in the same place.

"Great, ain't it?" he said, drawing in the fresher air as if it were wine. "There's a way out through them old workings, and we'll have some air in here from now on. Ought ter get the V.C., you did."

He was right about the ventilation, and they repaired an old main roadway back to the mouth which was hidden by a hollow in the other side of the mountain. It made a good return, and their antiquated fan was able to draw more air into the workings.

The neatness of those old workings made our places seem quite slovenly. Not a stone was laid out of place, the ribs of the coal were cut perfectly straight and strong, well-built walls made a deal of posting unnecessary. They had been worked in more leisurely times, when the rush of coal-getting was not so intense, even if the hours were longer.

The old-timers in the village were very interested in accounts of these old workings. One man, who was nearing his century of years, claimed that his father had carried him up there on his back when he was six years old and taken him inside every day. Although the child would not be able to help in the work, he would be sitting on the side all day, and his father would be able to claim the turn of trams that would be given to two people.

A few miles from here they opened some old workings that must have been worked about the same time as were those I discovered. These other workings had not been used since an explosion, and they found several skeletons in there that could only have been the remains of children of the ages of six or seven.

After this our management altered their methods of working the coal: they copied the methods of the old colliery and cut the bottom. This did not break the top so much, and it was very successful. I did not get the credit I thought my due from this adventure, for everyone wanted to pull the leg of the fellow who fell into the next world. Even the management, despite the benefits my fall had given to them, could not be persuaded that they ought to pay ripping price of twopence an inch for my advance—downwards.

One of my friends tried to take advantage of this other way out to sneak away early one Saturday when he was to play football. He waited for the fireman to pass, dodged into the old workings, and reached the open mountain through the old mouth. Once outside, he found that the top of the mountain was covered with a thick fog, and soon lost his way. He wandered about until he had no idea where he could be. Some time later he noticed a fowl walking along, and realising that the fowl had a home somewhere, he chased the poor fowl until it squawked back to the farm-house. He had gone to the left when he should have kept to the right, and was in the valley on the other side of the mountain. He was loaned money to return by train, and arrived home late that night. He had not been able to wash or change his working clothes.

Now there was a better supply of air in the workings the management wanted the price list settled without delay. There were nightly committee meetings and general meetings to settle what it was worth to stand posts, cut the bottom, rip the top, work in water, drive airways, fix cogs, notch double timber, and a score of other jobs. Most of these items were settled without much delay, but the greatest discussion was over the cutting price of coal.

The colliery had two districts—the lower district, which was called the Deeps, and the other side, which was called the Rise District. In the Deeps the coal worked by hand with a mandril, and, as the clod was thin, it was possible for a good collier to earn more than the minimum wage if he was paid at the rate of three and eightpence a ton. In the Rise District the coal was stiff; it had to be wedged free, or blown out with powder. Two men had to work very hard to fill three trams between them. It is a well-filled tram that holds a ton of large coal, although the total amount of large and small may be over thirty hundredweight. The collier gets nothing for the small coal, or even for the small lumps known as peas.

The concern of the coal-owners was to fix the price so that the men in the good places should not earn much more than the minimum wage; while the committee pointed out that the men had no chance to make their wages in the stiffer coal of the Rise District. The management wanted to fix the rate on the basis of the best places, and argued that the poor ones would improve; the committee wanted a chance of getting their wages for the men in the poor places, and argued that the good places were quite as likely to become stiff.

At the time when this argument was at its keenest phase our place finished, and Tommy and I were sent to work up the Rise. We did not want to go, but were compelled to—I have since thought there was an intention behind the move. We worked very hard during the first week, and filled no more coal than the men on either side of us. Towards mid-day on the Saturday we noticed that the holing in our working-place was getting quite soft. In this seam we holed above the coal to free the clod, then lifted the coal. To have stiff holing meant that we had to keep pounding away at the coal all day as if we were hitting a cement wall; in this new place of ours this holing suddenly became so soft that our mandrils took lumps out of it with each stroke.

Our coal, too, became as easily workable as the holing; yet on both sides of us men were battering their hands to pieces in a vain effort to do what we could accomplish with hardly any effort.

We were both young and strong. By mid-day on the following Monday we had three trams full, while no one else near us had their second. At that time the manager and the fireman came to us. They noted the number of trams we had filled, and we told them the coal was working better. I could see that the manager was excited over it. He was a keen man and saw possibilities. He asked how many trams a day we thought we would fill, and we said that if things kept as they were, we could manage five a day.

"I'll tell you what I'll do," he said, after a brief consideration; "if you'll slog in and get thirty-six good trams full between you—that's six every day for this week—I'll give each of you a pound bonus on top of what you earn."

We went all out for that pound each and the good name that the manager would give us. We had a devil of a time filling six every day; sometimes we worked later than the others by two hours, and we stole every bit of coal we could find on the sides or anywhere near us after the others had gone home.

By the Saturday afternoon of that week we did get our eighteen each full, and wanted no more. No one else in that district had twenty full between two workers, although some of them had worked very hard; for it was a strange thing that the softness in the coal only existed in the twelve yards that we worked forward.

We had our pound. It was added to our three pounds sixteen of wages. It was great pay for a week's work as counted in that area—four pounds sixteen. We went home very proud of our earning ability, and with the slogging that had been done quite forgotten.

What neither of us realised until afterwards was that for that pound we had betrayed all hope of our fellow-colliers having the highest price for cutting coal. Their claims were defeated, because it was shown that a large amount of coal could be filled in the Rise District, where the working was counted to be stiff. Our holing kept soft for about three weeks, then it went back stiff, and after that no one ever earned more than, or even the amount of, his minimum wage.

We realised, when it was too late, that we had been the cause of the rate being fixed so low that the men were always afterwards driven like slaves in an attempt to force them to earn as near their money as possible, while there was no hope of them getting a shilling over their rate.

My regret over the result of this episode and the knowledge that it was our youth and inexperience that had allowed us to make this mistake caused me to attend the general meetings and take more interest in the doings of our committee. These meetings were held in the concert-room of a public-house; there was no other building available then.

Sometimes a man who had a grievance to report would arrive early and have a few—or a lot of—drinks to make him eloquent when the time came to state his complaint. There were several Durham miners working there, and one in particular had a loud voice and a terrific accent. He would nearly always have a complaint ready for each meeting, and considered it necessary to be in the bar about an hour before the meeting started so that he could have a few drinks and work himself up into a temper about the grievance, whatever it was.

Often the meeting had started before he would consider himself in the right condition to attack, and when he went he made sure of taking some liquid reviver with him. Perhaps right in the middle of the secretary's reading of the minutes this miner would crash in through the doorway holding a pint mug of beer in his hand, and demand in a voice that would silence everyone else:—

" Naw! Aht with it. What abaht that what ah tole to thee, eh? "

Before anyone had time to recover he would march up the centre, demanding as he went:—

" Answer me that, tha' gret gowk."

He would not pause until he was near the chairman's seat and was annoying that man by waving a partly full mug of beer under his nose while he shouted:—

" Say summat, tha' gret gowk, if tha's not afeared. Tha' dursent. Tha's afeared."

Then he would empty the mug with a loud and long gurgling, and wave his free hand toward the Lodge officials to emphasise his statement of:—

" Tha' sees? They'm a bunch o' bloody twisters. They are, an' all."

Back down the centre he would march then, only pausing before he slammed the door behind himself to inform the grinning crowd:—

"An' tha's as bad. The whole lot on thee."

There was another who always sat near the door. When he felt compelled to say something he would get up to his feet and half open the door. Then he would spurt out a string of eager, rapid words which were practically rushed into one another. When he had reached the end—it never lasted many seconds—he stepped outside the door and slammed it. We would see no more of him that night.

What a mixture of languages and dialects were there sometimes! Yorkshire and Durham men, Londoners, men from the Forest of Dean, North Welshmen—whose language is much deeper and more pure than the others from South Wales—two Australians, four Frenchmen, and several coloured gentlemen.

Of course, the Welshmen were at a disadvantage when they tried to convey their thoughts in the mixture of languages that is called English. The meetings had to be in English because most of the Welshmen could express themselves to some extent in English, while the majority of the English maintained a frightened silence whenever Welsh was spoken.

We had the man who insisted on translating a Welsh speech into English or an English speech into Welsh. He was the one who suggested "that we draws this meeting to a shut"; and who later on disagreed with "these stay-stop strikes" when he meant a stay-in strike. I remember one of the Frenchmen asking a question of the miners' agent. It was phrased queerly, so the agent suggested—rather pompously—that he could understand the speaker better if he spoke in his proper language—which he thought must be Welsh. The questioner used his own language, very readily, to the complete dismay of the agent, who had to confess that he had better try it in English again.

CHAPTER SEVEN

We had moved to our apartments—two rooms. I remember how proud I felt when I saw them furnished for the first time, and realised that all that shining furniture was ours—even if most of it still had to be paid for—and that the room was mine to sit down where I wished. The novelty of it kept our thoughts concerning the drawbacks quiet for some weeks.

The front room, our living-room, was about ten feet square, and the bedroom about the same size. Our words had to be low spoken, else they could be heard in the kitchen. The scrape of a chair, or even the creak of a bed, could be heard by the other family. Certainly they had only one child, and we none at that time, so we were not so crowded or noisy as those other houses—and they were many—where a considerable family was living in the front and back of each house.

All our work had to be done in that one room. It was necessary to have a large sack to cover the mats when I came in from work, and it had to be kept down until I had bathed and the heavy tub had been carried out through the kitchen to the back. My working clothes had to be packed into a large box and pushed under the table.

I remember there was only a toy of a grate in our room. My wife wanted to try her hand at different sorts of cooking, but she had to wait to use the oven until the people in the kitchen were agreeable for her to do so, and then it depended on whether they had cleaned under it that morning so that it could warm. It was only natural that they should not always welcome someone walking in and out of their room. Sometimes, I suspect, the wife did not want her husband to see that someone else was cooking and she was not in a position to do so.

Washing the clothes was a problem. All the water had to be carried from the street taps along the polished oilcloth, and that caused black looks. Often there was only one day a week on which the clothes could be put on the line—the day when the wind was blowing up the valley and the dust from the colliery-screens did not come our way. On that day the people in the kitchen would want their washing out. They claimed the strip of garden, and it had only room for one line, so our washing was done in the room and the drying on lines under the ceiling. My dressing and undressing from dusty working clothes did not improve the clothes or the furniture.

The colliery company built its houses as close together as possible, but reversed this method when placing poles to hold the electric lamps. They averaged three lamps to each long street. They did not always allow the householder to say who should come in apartments with him, but they had a way of making sure he would take someone.

Rent and electric light as well as coal were stopped at the office out of the pay—and they were a first charge. If a man had twenty-five shillings to draw, and his rent, coal, and light came to twenty-four and sixpence, then he went home with sixpence. There are cases when only a boy is working in a household where the father is injured or dead; and at the end of the week he has not enough pay to admit him to the pictures.

We had to go through the rooms of the other people for anything we wanted, even to go to the lavatories. These were the only relief to the grey stones of the streets; they were of red brick, built at the farthest end of the garden, and were placed in a high position for everyone to admire.

I had my share of bumps and bad luck at the start of married life. Most miners seem to get the same during the first year: just the period when they are most anxious to work full.

One day—it had been a terribly wet morning, and there were many missing from work—I got caught by a fall of clod and coal. This fall knocked the light out of my lamp, and I was held fast by the coal. I think that I had only bruised my leg then, but I was in the dark, and no help was to be seen. Also the top was cracking above me as if more was about to fall. I struggled so much to get free that I wrenched my right knee quite out of its joint.

I failed to get free, so had to wait some time until a haulier arrived and got help to release me. I pretended not to be much hurt, because I had a lot of coal ready to fill, and I insisted on keeping at work—how I did so has puzzled me ever since. When finishing time came I could not put my foot to the ground, and had to be carried out after a few vain attempts to walk. Outside the level they took a horse from its work on the tip and lifted me on its back. It was an uncomfortable ride, and when they passed the gate on the mountain they did not open it wide enough, so my injured leg bumped against it and I was emphatic about it.

The doctor was in the street when they reached my home, and he followed us inside. While they were exposing my leg he was examining the pictures on the wall. He did not like one of them, and said so. Then he saw a pile of sliced potatoes that my wife had been preparing for frying. He thought they were sliced apples and commented:—

"I like a nice sliced apple myself."

He put several pieces in his mouth. What a face he pulled when he started to crunch them!

That doctor was a character. He was a short, slight man, a North Welshman, who had come there as a young man forty years before. He knew the people of that village—the whole five thousand—as a mother knows her children. When he thought there was need, he could be a first-class doctor; when he did not feel that way, they could do what they liked about it. He had a habit of coming in and criticising the furniture or the pictures, and his remarks would start an argument which might cause the reason of his visit to be forgotten. I remember him going away in a temper after a heated debate and returning five minutes later to inquire:—

"Oh yes. I've come back to ask who is ill here?"

On the rare occasions that he had time to spare he was fond of paying calls just to see how people were living. Sometimes they would not be in the house; then he left his mark by placing a chair on the table. The children used to depart at top speed when they saw him coming, for he had a wicked twinkle in his eye and a way of carrying his walking-stick hidden behind his back. I remember his method of keeping the children quiet in the surgery:—

"Now, the next child that speaks will have a tooth pulled out by me."

What a silence there was after that!

His surgery was a wooden-sided, zinc-roofed structure that was divided into three. In the outside part the patients waited, and those who were first got seats on the two wooden benches, and could see what was the smallest amount of fire it was possible to keep alight. In the middle part the dispenser shivered and muttered at his fireless state with only the coloured bottles to warm him; and in the farthest room the doctor sat near a good fire, quite unconcerned about who waited, as long as he had someone with him who was interesting.

This surgery was lighted by oil lamps hooked on the side. They were the single-burner type, one in each room. When the glass got broken they were stuck together with sticking-plaster, until some were more sticking-plaster than glass. It was not safe to clean them in that condition, so the light they gave was dimmed. A woman went to turn the wick a little higher one night, and the glass fell to pieces. The doctor came out, then began to lecture her for interfering with his lamps. She was a big woman with a corresponding voice, and she was no slouch at answering back. After she had silenced the doctor she went out, but returned soon after with a lamp-glass, which she placed in position slowly, watched by about thirty patients. Then she shouted in to the doctor :—

"There you are, you old b——. You won't need to buy a new glass, so you can sleep to-night."

He drew twopence in the pound earned from every man that worked in that area, and there were several collieries there. He must have had their panel money as well. He left a very large sum when he died.

I was in the house for nine weeks with that injury, then re-started work before I should have, my knee beïng still weak. I was paid compensation, but the amount and the method of paying it did not help me to get better. They gave the impression that I was trying to get something for nothing out of them. This compensation was handled by an insurance company, and all the injured were treated as if they had crippled themselves deliberately.

I had to hobble the distance to the office several times before I had any payment. I think I waited nearly a month before they started to pay anything. In any case, they are the same as the sick clubs and the Health Insurance and the Dole: they must think that the working people have so much money that they do not miss the first three days. They cannot seem to realise that most of us are living on our next week's wages, and that even a day out of that means that someone must go short.

What a compensation it was when it was paid! They had a way of assessing our earnings and paying an amount that was roughly about half what I would have earned if working. So to add to my suffering I had the misery of knowing that we were going behind financially; therefore I could not afford to stay in the house until I had recovered.

On the last shift of the first week after re-starting I was injured again on the same knee. A piece of clod rolled out and hit me. That caused me to have another seven weeks at home, and the insurance people were quite convinced that I enjoyed getting knocked about. This injury happened on the day before some Bank Holidays. I often wonder if we are justified in calling them holidays? It is grand to be relieved from danger and to be able to walk the mountains, to hear the birds, and feel the sun, but pay-day always comes, and that one will be days short. We have days off at Easter, then, before the payments have been squared again, it is Whitsun, with more lost days. Of course, we never know when slack time may come, with continual stoppages.

There had seemed a prospect that the strike at the colliery where my father-in-law was employed would be settled and that the men would gain their point. I remember that he came home elated one night with the news that things would soon be better.

On the following Monday morning a train came up the valley from another district, and the rumour went around that a crowd of men had been brought to work that colliery as blacklegs. It was correct, and about eighty men were brought, with about half their number of police. They came down day after day, and the police guarded them while they went to work. They were not allowed inside the Miners' Federation, and their coal was weighed and checked by the man the company favoured. The miners from Treclewyd marched there and watched the blacklegs going to their work. When weeks had passed, and the strangers still came, the Treclewyd men began to see that this outside labour would cause the strike to be a failure, and some of them tried to get their work back.

Some were re-engaged; but none that had been prominent in the Union activities before or during the strike. My father-in-law was prevented by police from going on the colliery, as was the other check-weigher whom the men had newly elected. The man they had tried to sack was guarded by police to weigh the coal in a way that pleased the company. That was how the men got their coal weighed independently at that colliery.

My father-in-law could get no work at all in that area for nearly five years. Even then it was only as a surface labourer at another colliery. The agent got him that work on condition, so I understand, that he took no active part in the Trade-Union activities. His health—and his feelings, I suspect—

did not stand the strain of that work long; he failed altogether.

My mother was ill about that time, and I went home to see her. I went alone, leaving my wife with a newly born daughter—Rose. While I was in Hereford, Germany started to march. I listened in the schoolroom near home to an aged, but very savage, colonel explaining why we ought to annihilate Germany, and a patient vicar proving how on our side this was a just war—a most Christian war. I was impressed. The next day I saw a cousin of mine in khaki.

This cousin—much older than I was—had always desired to be a policeman. He was well built and smart, but lacked one quarter of an inch in height. He had worn out several developers and stretched his legs to their limit, without success. He had joined the fire brigade; had shouted a lot at fires; he had lifted his hat to police inspectors' wives; had spoken authoritatively to police sergeants about gardening and the shameful ways of criminals; had accompanied constables for walks and adopted their swing of foot and shoulder—as well as way of speaking—but he could not persuade himself to be any taller than five feet nine and three-quarters. He had been in the Volunteers for years, had been made a sergeant, and had got a medal for long service.

Then the War came and he was made a Military Policeman. It was a wonderful opportunity. On the morning before I was to return to Treclewyd, he clamped along the pavement and saw me.

"Hallo!" he said, and his manner of speech had gone much rougher during those last few days; "what's this? You ought to be out in France, or getting ready to go."

" Oh yes," I could not help replying. " Then how is it you're not out there? "

" Me? " he snorted. " Don't forget I'm a married man, and have got my wife and child to consider."

I was so impressed by his swagger and the speeches at the meeting that I walked back and forth several times past the recruiting office, and at last followed another man in. I planned to get in the Hussars, but was rejected—to my surprise. Possibly the unnatural lighting in my work or the amount of reading I had done, had affected my eyes. Anyway, the card was a complete blur, and I began to realise that my eyes were not what they had been. It was only after I came away from that office that I remembered that I, too, had a wife and child.

My mother recovered, and I went back to work. Most of the men who had been idle through the strike became employed at our colliery, for it was opening out work very fast. One who came there was James Davies. He was small, quiet, and as inoffensive as a man could be. But he brought with him a reputation that stopped the boys from teasing him and made the officials avoid quarrelling with him.

One day, while working at the other colliery, James had been cutting a post with his hatchet. A fireman had spoken to James. He had been busy and did not hear. The official repeated his question and embroidered it. James, suddenly aware that the official was shouting at him, straightened up and walked across. He still held the hatchet in his hand—quite a reasonable thing to do, because to lay it down anywhere might mean hitting the sharpened edge against a stone, apart from the danger of leaving an edged tool lying in the dark roadway. It may have been that the nerves of that official were upset or his conscience warned him of what his behaviour deserved; anyhow, he saw in the advancing James an avenger chasing him with a hatchet. He fled swiftly, shouting the alarm as he went that a madman was chasing him.

James, startled in his turn, was tackled roughly from behind, robbed of his hatchet despite his protests that he wanted to finish that post, and taken outside, where the official's story was believed, and James was sacked. The magistrates, too, believed the official, and fined James a total of five pounds. So poor James, who had been careful not to slice a worm when digging, stayed away from chapel because of his shame, and when he did get work again after a long spell of unemployment, he went about that work with the noise of the shyest mouse. But no officials argued with him, no young boys plagued him, and on the rare occasions when he did make a statement he was listened to with respect, for was there not the danger that if he was annoyed he might once again get one of those fits and go chasing people with a hatchet in his hand and the lust for blood in his heart?

I had one collier working next to me who was an expert at counting his money before it was earned. He classed me as a better scholar than himself, so wanted me to count up his total before the measurers arrived. So many yards of ripping; so many posts; so much double timber; half a turn extra for this and another for that—he was quite good at making it come to a decent total. I usually counted up to about four pounds, and that would satisfy him.

"That'll be about right," he would agree—"m'yes, that'll be about it." There was always a row when the measurers went into his stall, and after they had gone he would come up to me and complain, " Them sods have robbed me of more'n thirty bob this week agen."

Another liked to do his work about a fortnight in advance. If he walked home with me he would be planning all the way:—

" Holiday pay next week, ain't it? "

When I agreed with him, he would continue:—

" Must get a bit extra out for that. If I fills twenty-five trams and rips my bottom on crop to the coal, and stands three pair of timber and about five posts, and cuts them puckings back on the road, I oughter get about three pun ten, especially if they gives me a bit of allowance for the water, as they ought, eh? Could take the missus and the kids to Swansea for the day, then."

Poor old Bob Jones! he nearly always found it far easier to talk about work than do it.

There were two men who were always conspiring as to how they were going to dodge a bit more pocket-money from their pay. Their wives saw the pay-dockets and had learned to understand them; also they seemed to keep a firm grip on the money. These men were often adding little items to the list of deductions and taking them from the total. One had an item of two and sixpence for rails that he had to buy. This was added in ink and deducted from the total. The wife did not suspect, so next week he had to buy a cross-parting at three shillings—remarkable bargain—and the week after a wheel-barrow was needed to wheel the coal back. This one was dirt cheap at three shillings too, but the continual extras aroused the indignation of the wife, and she protested strongly to one of the clerks in the street. After that there were no more extras in ink on the bottom of the docket.

We were desperately short of materials to work

with, and delays in getting coal back were frequent. Rails were very scarce. We had to trim slight-posts sometimes, and lay them down for the trams to pass over, and broken water-pipes were used as well. Often we had to lift a rail up when the tram had passed and hurry in front of the haulier so that we could lay it again farther on. Trams, too, were scarce in the first years of the War, as they could not be replaced when smashed.

Often we spent hours with coal ready to fill and no tram to put it into. So scarce were trams at one time that the haulier used to bring us one and wait for us to fill it, so that he could take it out again, as there were no others for him to handle.

It seemed strange, after being idle most of our time during the working week, for the officials to ask us to work on during the Saturday afternoon to fill coal—this we did on several occasions.

Our company did seem to make every effort to give us trams when they could. I know another firm—a much bigger one—where the colliers were waiting day after day for the trams that never came. Some days they had no need to take their tools from the locking-bar. The week's output between two men was never more than six, when they could easily have filled forty. I suppose the subsidy covered the costs.

One effect of this slackness was that after the War the colliery was closed as unprofitable. This action was based on the costs during that period, when the men could not get the trams to fill, although they wanted them.

If engines and material of that sort were scarce, there seemed no shortage of bricks, and fine engine-houses were built to shelter engines that were little better than scrap iron; imposing bridges were built where wooden gantries had been before; and several companies built large villas for their over-officials. I suppose it is a good plan to prepare for the future when one has a generous banker.

So many men had joined the Army from our colliery that notices were posted up telling the men that they were of more use to the country if they stayed at work, and there was a time when it seemed we should have to stop working through shortage of miners. Some were sent back who had joined the Army, and more than fifty came from near Swansea to work with us because their pit was flooded out.

Belgian refugees came, too, adding another to the many nations that we had represented here. We had several rises in pay, but we only gained them because of the rise in the cost of living, and the price had very often been up for several weeks before we had an increase to offset it. Very often the shop-keepers added another little increase after we had our higher wages, so that they got us both ways.

When food was restricted nearly every one of us knew what the other was bringing to work—bread and treacle. No longer was there a shout of, "Coming to have food?" The new call was, "Coming to have that bread and treacle?"

Gradually they were introducing the double shift into the collieries, and the men did not like it. When only one shift is working it leaves time for the workings to be freed of the powder-smoke and the warm air that comes from the breath of a lot of men working at high speed. The workings can rest and settle for a period, and as there is always a moving pressure from the roof on to the edge of the unworked coal, it happens sometimes that there is a large amount of coal that is squeezed free by the time that a man returns to work in the morning. An experienced collier will help this loosening along by trimming and cleaning the slips of coal, so that they can slacken and give him easy coal for the morrow—for the slips of coal lie on one another like the leaves of a book, and must be lifted one by one. The careful workman will put his timbers where they are needed, and will see that they are solid and well sloped. When the double shift is worked, the good man must suffer if a poor workman is sent to work on the other shift—and the officials often arrange it that way. His loose coal will be taken, and only a dry face of coal left; timber may not be placed at all, or may be badly placed and, as such, be a death-trap.

Often it is the careful workman that is injured, and by some neglect of the less careful one, who always seems to escape himself. There is the method, too, of forcing one shift to beat the other at output, and this can often only be done by neglecting the timbering. The careful man insists on putting the place safe, so he gets behind on his coal output, is reprimanded, and goes short in his pay, or is sacked. There is a small additional payment per ton to compensate for this double shift, but it does not nearly cover the inconvenience and drawbacks.

After the double shift was going properly, rumours started that they were going to try coal-cutting machines in that colliery. These machines were rare in South Wales then, and there were none working in our valley or anywhere quite near. Tommy and I were set to open a cutting face for this machine. It had to be about one hundred and fifty yards long, and as straight as possible. There was a commotion over that machine-face; several times a day we had to put our lamps in a row so that the officials could survey which was behind and which in front. There was no question of working only eight hours at our place then; one man who wanted to go away and would not work on was sacked.

If Tommy would not get to work in time in the morning, he made up for it at finishing time. He was never anxious to go out from work. He always had something he would do after all the others had gone, and it mostly took an hour or more to complete. Then, as we tramped over the mountain together, he would tell me of his ambitions.

Tommy did not intend being a miner for long, and he wanted to make as much money from it as he could in a short time. He was trying to save quickly, for there was a little cottage at Blackpill that he wanted very badly, because it was handy for watching the fishing-nets, and he was thinking of getting married.

He succeeded in the marrying part of this programme. She was a girl from that village. I remember that we worked the Friday together, and so late into the Friday afternoon that we only had time to tramp home, get more food, see that our money had been paid properly, and return to work. There was not even time to bath, for we had to be back across the mountain by eleven that night and ready for the night-shift.

We had twenty-four hours of hard slogging straight off, and Tommy was getting married on the Saturday morning. That was no way to treat the marriage ceremony, nor a young bride. I was too tired to go to the house to see him off, so I got a chair by our front door and waited to wave when he went by. I was sound asleep sitting up when they passed, and never saw them, but it did not matter, for Tommy himself was fast asleep in the car, and they told me that they could not keep him awake long enough to have their photographs taken.

During the winter it was a rough journey over that

mountain in the night. We had a black mountain to cross, with no lights to show us the way, for even our working lamps could not be kept alight in the fierce wind that was up there. The streams were often flooded, and we had to wade through them by guess, and sometimes up to our knees in water. Often we would stand on the side of the biggest brook, listening to the rush of the water, and feeling about in the darkness for the plank that was the only bridge. We had gantries that bridged gullies of fifty feet in depth to cross in the pitch darkness, with the danger of going right down if we did not stride the exact distance to the next foothold. Sometimes for several days and nights the whole mountain was a sheet of ice, over which we crawled upwards on hands and knees. It was nearly always on the most severe nights that the accidents happened and we had to carry men home. Often another man had to walk alongside each stretcher-bearer to support him in case he slipped, and to guide his feet to the safer walking. Sometimes it took us three hours to get an injured man through the bitter night. First-aid requirements were neglected there too, and we had to use our own overcoats to cover the injured, and often our own scarves to bind them up. There was a dose of husky speaking and bad colds after each journey of that kind; while we went down the mountain in fear that the old stretcher might crack at any moment.

On other nights, after we had splashed through the streams and bumped into trams and stones on the way up, we would reach that warm cabin on the crest of the mountain only to discover that it was too rough for any of the men to stand out on the tip to unload the rubbish. This meant that we could have no empty trams, so back home we had to struggle, arriving there about one o'clock in the morning and having to bath all over again. We had no pay of any sort for these little pleasure-trips, and I have known such to happen twice in the same week.

The man who had his lamp flattened started for work one morning, and when half-way there he met a bull. He should have been able to defy bulls, after what he endured at home, but for some reason he was very much afraid. This bull did not attempt to hurt him, only to stop him going past and on to work. Probably the bull was lonely; anyway, he was friendly until the man tried to get past, then he got angry. Between abortive attempts to pass and quick retreats when the bull seemed annoyed, the man never reached work that morning, and the bull escorted him back almost to his home. The boys were fond of making a noise like a bull behind him after that episode, and they advised him to go to Spain, where all the ladies would be sure to love him for his daring bull-fighting. Someone sympathised with him once because of his wife's treatment.

" But she don't boss me about, mun," he objected —" she do only keep me in my place."

So it seems he was content.

While we had been preparing the long coal-face, things had been coming into the colliery yard: thick electric cables and bigger switch-boxes, as well as other parts of machines that we had never seen the like of before. We examined them well on Sunday afternoons, and made vague but hopeful attempts to explain their construction and use.

Then the expert arrived. He was a very energetic fellow, with a suit of overalls and a Yankee accent. He knew how to fit those complicated parts together, for he was the staff man from the coal-cutter works, and had only recently come across from America; but he did not know how to carry a lamp, so someone had to walk alongside him and show him light. The next day he arrived near where I was at work, and the parts of the machine were there before him. After a lot of loud shouting and heavy lifting, we got it fitted together. Then a long, rubber, snake-like thing was trailed down the coal-face and we were warned:

"Say, you fellows. If you're wise you'll keep away from that thar cable. Thar's enough electric juice in thar to shrivel the hides of on yuh."

Having no wish to shrivel just then, we obeyed. The majority obeyed to the extreme, and kept as far away as possible, but the hum of an engine has always been music to my ears, and this powerful machine interested me very much. I watched his every move, and asked a lot of questions. His bluster was only the veneer over a decent instinct, and he showed me very willingly all he could.

I recall the first day that he demonstrated it. The one who was to hold his lamp had gone away to get something, and I had stayed to help him with a jack. My own work was waiting, and so I turned to go back to it, not thinking that he had not a light of his own, as we all had usually.

I jumped when a voice bellowed behind me:—

"Great Shucks!"—it sounded somewhat like that—"where the h—— d'you reckon yuh are going with that flicker, eh? D'you reckon I'm a b—— owl, or what?"

I was in a bad humour, and replied that he sounded like an owl, anyway.

"Oh! Ah do, do I? Let me tell yuh, me lad, that if ah was a few years younger ah'd show yuh whether I was an owl or not," he bawled back.

He was young enough—under forty, I should think—to make this statement seem ridiculous, and so I laughed and replied:—

"You mean, I suppose, that you'd try."

He glared at me for some seconds, then laughed heartily.

"Come on, Taffy," he miscalled me, "show me a bit of flicker here, for the love o' Mike, or this blasted machine won't start to cut to-day."

It was a powerful machine, and seemed immense in comparison to the limited space it had to work in. It was over ten feet long to the extent of the jib, eighteen inches high, and over two feet wide. It weighed three and a half tons, and was so heavily steel plated that no falls of roof would damage it. It had a hundred yards of electric cable; thick cable this and flexible; always called the running-cable, but I never saw it run much—in fact, it was very hard work dragging it along the coal-face.

The chain had thirty-eight detachable picks screwed into it, and revolved at about three hundred revolutions a minute around the four-feet-nine-long jib. It undercut the coal to the depth of the jib—four feet nine—and travelled forward at about a yard a minute, being drawn by a steel rope fitted to a sloping post and moving along on a steel slide. It made a terrific noise when cutting, and the dust was enough to choke anyone very close by. It was a waste of effort to speak when near it—the only way to explain was to wave your lamp and make signs.

What a collier could do by working hard all day, this machine would accomplish in two minutes. The holing had been such hard and monotonous work that I considered this machine a boon, for it did the very hard slogging for us. The drawback was in the added danger, because we could not hear the roof cracking, and with such a large undercut there was the likelihood of it falling any second. Timber could not be set so close, either, because space had to be left for the passage of the machine.

Some of the men took a long time to learn that they had to keep their working-place straight and clear of posts. They seemed to imagine that this powerful machine was made of elastic, and could be turned at any angle.

If a post or stone gob was left in the way, the machine would shift it all right, but there might be hundreds of tons above that was steadied by the gob or post, and that would come down in the shifting.

A man who had worked with machines in the North of England had been persuaded to come there to work the machine. He had been offered more money than he was earning elsewhere, but was soon regretting the change. It was a type of machine that he had never before seen, and the seam was strange; he could not become accustomed to it.

The coal-cutter seemed to dislike him too, for it would stop working on almost every occasion that he was trying to handle it. I remember watching him on the sly one day when the machine was stuck and he thought no one was about. He took some stones from the gob and slammed them at the steel sides of the machine. As each stone hit it he growled:—

"Now go, you Yankee b——, you."

It made no difference; his luck was out, and he was disheartened. He went back twice with the excuse that the noise made him ill, then got drunk a few times and was ill in reality. He begged to be taken off the job, and the demonstrator wanted to go back to the firm, but had to get the machine going properly before he did so.

He suggested to the manager that he try to persuade " that young feller that was with me the first day. He's the likeliest coon for the job that I've set me eyes on in this area so far."

I refused at first, for the work was reputed to be very dangerous, even among men who had spent a lifetime in danger, but the throb of the machine was in the pulse of my blood, and I agreed at last—after I had been guaranteed a day wage above the minimum rate and a lot of overtime. The American gave me a booklet, explained the workings of the machine and the electric circuit, then I had a trial run, during which he stated that I handled it almost as well as he did himself.

" Born for the job," was his comment, " and we've bin wasting time with that other goosoon."

The manager warned me to be careful how I handled it, because it was worth well over a thousand pounds, yet the minute after he told me he wanted that machine pushed for all I was worth, because the output would very soon all depend on it, for they were abandoning hand-cutting.

I was given a blue paper vesting authority for the handling of machinery and electricity, and found myself a coal-cutter operator with the sincere, if profane, good wishes of the departing American.

I was given the chance to pick my mates for the work, and although Tommy did not like the new job, he preferred doing it to parting with me, so he came. His job was posting behind the machine, and he was helped by another young chap whose name was shortened to Hutch for our convenience, and who was never called anything else. He was a good worker, and not afraid of anything under the ground. He was well built, too, and used to fancy himself no end. He was living at some distance in the other direction from our village, and I think he must have told the people there that he was an important person at the colliery.

He used to wear a very good overcoat and trilby hat up to work, and leave them very carefully behind the boilers. After we had finished the shift he would undo the straps from his knees, dust his trousers and boots, take off the dirty jacket and cap, and put on a cleaner jacket, the belted coat, and the trilby. He had quite a swagger after that, and went down the incline looking more important than did the manager. We were good friends, all the same.

The fourth member of the working gang was George Bennett. He came from the borders of Herefordshire, and was a year older than myself. We were all in the early twenties, and were but little concerned with the risks we were running. Every hour of the day I was learning things about that machine, and in spare hours I used to study the diagrams in the booklet. I had to find out from experience, and that was often painful. I learned to respect electricity—but only after I had been knocked down by it twice and had days when my wrists and ankles were made useless by the shocks.

Stones fell from the roof and sliced through the outer casing of that thick cable; then there came a flash of flame and the switches blew out. I had my waistcoat partly burnt while on my back on one occasion, and Hutch's trousers were singed by the current that was leaking along the rope. Picks wedged in the chain, and we had to cut out sections to replace them; to splice the rope that drew the machine; to adjust the sprocket springs on the haulage gear; to replace blow-out switches, and to see that the score of holes for the oil and grease were working as they should. On one occasion we were on stop, and thought the electric fingers that touched the controller were to blame. We opened the machine and rubbed the faulty finger with a rough stone, then with the striking side of a matchbox, because we had no sand-paper. When the chain started to fly round again and the roar assured us that we had repaired the fault, we shook hands with one another very gravely to show our appreciation of our own skill and ingenuity.

The mechanic and the electrician were only there on the day-shift. It would have taken more than two hours to get them there after that, so we had to manage, and they knew little more about this strange machine than we did; but we were using it for quite sixteen hours of each day, so we were learning faster than they were, and we were accustomed to being underground, while they dreaded and detested coming there at all.

Very soon the management started to pit the machine against the men. There was a speeding up in all the jobs. When the demonstrator had started this machine, the manager had been delighted with a cut of a hundred yards long in the eight hours. Cutting is only part of the work—oiling and greasing must be done constantly, picks have to be changed, timber has to be kept tight up, and the machine has to be guided and stayered to keep alongside the coal. We learned how to do these things quickly, and we were a gang that worked to answer one another very well, so that in little more than a month we managed to cut a hundred and sixty yards in a shift. The officials made a fuss about this new record, and professed to be proud of us, but they were very sarcastic if we cut any less in a shift afterwards.

It was necessary to wear leather knee-pads because we wore through thick trousers in one shift, and our knees were stiff even after a rest at the week-end. About this period the week-end break lasted from two o'clock on the Saturday until six o'clock on Sunday morning. We never knew when we would finish work. Every day was an emergency. The officials would be behind the colliers all day hurrying them to clear the coal, else they would delay the machine, then they would hurry back to warn us to hurry up, or the colliers would be clear and waiting for coal. That was the way of it every day —rush us up to the colliers, and rush them so that they should not delay us.

Soon the cutting-face was extended to almost five hundred yards, and we always had a stretch waiting for us. For months we worked a double shift each day, crawling on our knees under low height for sixteen busy hours. Several times I went to work on the Sunday morning and returned home late on the Monday afternoon. One week I was in bed only eight hours in all. We had to start for work at various hours, whenever it suited the management. We went to call for Tommy one evening, and his wife said that she had failed to wake him, so she asked us to try. We took off our working boots and went upstairs. We shook him roughly, but he only grunted; we lifted him up and punched him, but he did not mind. Then we carried him from the bed on to the oilcloth so that we could pummel him easier; but it made no impression, he still snored away, and so we had to put him back into bed and leave him to that glorious sleep—with very sincere regrets on our side.

On another occasion it started to rain while we were going homeward across the mountain at about two o'clock in the morning. We turned from the incline and sheltered in a stone cabin that was used by a man who kept the incline in repair. We sat down on some oil cans and waited for the shower to pass so that we could go on. When he opened his cabin door, nearly five hours after, the man found the four of us still fast asleep, with our wet clothes frozen on our bodies. Often I lay on the couch after bathing on the Saturday and my wife had difficulty in getting me to wake sufficiently to stumble up to bed at midnight. Twice I went to sleep with a cup of tea lifted to my mouth—with disastrous effect to the cups.

Despite the roar of the machine and the danger, I have dozed at my work very many times. I was careful always to crouch on the part of the machine nearest the waste-gob, and would be brought back to wakefulness by the machine pushing me gently along, and would find that my right hand still held the haulage lever in control while my left strained the electric cable taut so that it should not get underneath. There were times when I felt that it would not matter a great deal if the roof did fall on me, as long as the end of that torture to keep awake came swiftly.

Tommy and Hutch were in a cloud of thick dust all the hours, and often larger pieces of the coal were hooked out and shot past them or hit them. They had a sign language to one another like the deaf-and-dumb alphabet.

The revolving chain, with its sharp picks, was very dangerous. Sometimes these picks would meet a piece of iron stone and the whole machine would jump back. Anything or anybody that got close enough to those picks would be cut to pieces, for even if I switched off instantly, the speed of travelling would force the chain to revolve many times before it stopped. Hutch got his leg too near once, but his trousers were ripe so that they did not resist the pull of the chain—the bottom of the trouser-leg went and left his leg undamaged.

Very often a post was pushed too close—they were posts as thick as a man's thigh—but the machine would scarcely falter in its movement: it would give a sort of heavy grunt, then all we would see of the timber was a few thin shavings flying round on the points of the picks, while the cutter resumed its normal roar as if well pleased with the feed it had been given. A steel shovel was no trouble to it at all: the chain just took it, and we never saw even a square inch of it again. That meant another eighteen pence out of our pockets, and it happened as often as once a week; but even at that cost it was cheaper than hanging on to the tool and going into the chain with it. I have known of men being properly caught by these machines, but it was some years before it happened, and I hope I shall never have to experience that again.

I think it is a fair estimate to state that for every yard we cut along the coal-face a ton of large coal was mined. We had been steadily increasing our cutting; they had an indicator in the office to show how many minutes we were when changing picks and cables, and what time we started and finished. We used every possible way of increasing our cutting speed, even to eating our food in our hands as we worked, and at last brought the length of face we cut up to over two hundred yards in a shift. They had a check on the amount of coal we cut—it was over four hundred thousand tons in the first three years.

Where two men had filled three trams before, it

was now possible for a man and a boy to fill eight trams between them. Until the machine had been working, the coal was so stiff that the company had to make all the men up to the minimum wage. With the loosened coal of the machine the men could earn very good wages if the old rate was paid. It was paid for four weeks, and the men filled for all they were able—from ten to twelve tons a day. Then, when they had decided what was the limit a man could fill under these new conditions, the company insisted on reducing the rate, and they gave the men notice to finish the old agreement.

Assessing the costs of the machine-men's work, all running costs, and depreciation, it was admitted that sevenpence a ton would cover the cost of the machine-cutting. The men were agreeable to a reduction of that amount on the cutting price, and pointed out that the owners would not then have to make their wages up to the legal amount, as they had in the past.

The owners would not consider that offer, and said they would close down—the usual threat—if they did not get their terms. They were in a strong position, because the little repairs that must be done during a stoppage will be quickly offset when a reduction has been gained, and in any case they will be included in the cost of production.

The men were in a far different position. They have no money in reserve and are not allowed a quarter, or a half year, to get their bills paid. Their week's wage is often booked before they get it, and has to be spent entirely on the day it is received. Next week's food must come from somewhere, and rent, and clothes—such as it is possible for them to have. They have little chance to clear off debts, even if they could get someone to trust them. When they do restart after a stoppage they nearly always get a reduction in wages, so it is only natural that they should dread a stoppage and avoid it as long as they can. They are nearly always forced to listen to the doctrine that "half a loaf is better than none", no matter how much their natures rebel against it.

In the end the men gave way, and the new cutting price was one and eightpence a ton. This gave the company a chance to recoup their outlay very quickly, and ensured that the men would have to be extremely busy to earn their money.

It seemed that machine-mining would be a success there. The clod, which had been melting down when worked by the slower hand method, was getting drier and solider now that the coal was being worked so much faster—the coal-face was advancing four feet nine inches every working day.

The men became used to the machine to some extent, but few of them cared to be near it when it worked. In the wet places the current leaked through the damaged parts of the cable, and the air seemed to be full of electricity. We had the acid taste of it in our mouths for hours after we reached home. We were shocked into knowing where the cable leaked, and often placed a mandril near the weak part, with the sharpened blade just touching the leaking part. Then we would ask some collier working near to hand the mandril to us. His remarks after he had grabbed the mandril blade and received a shock were never anywhere near affectionate.

One man who was sitting across the cable happened to get on a live spot, and jumped up until his head hit the top six inches above him. He sat down again quickly, then jumped again and groaned.

"What's the matter, Fred?" we asked, very solicitously, although we knew he was sitting on a leaky part.

"Don't know, mun." He was completely puzzled. "I 'spects as it's them old rheumatics agen. Every time I puts my weight down I do get it summat awful."

George got in front of us one morning when we were going home—it was about half-past one, and very dark. He climbed into the graveyard of an old mountain chapel, then waited until we came. He gauged his time carefully so that he should spring out just as we passed. He was poising himself ready when he heard a sound right behind, and, turning in alarm, saw two eyes staring at him from alongside a white tombstone. We were startled when a speechless man tumbled over the wall and dropped near our feet, but we were calm in comparison to George. We found, after cautious investigation, that it was a mountain sheep grazing in the churchyard that had upset George's joke.

George decided that if he could learn to speak Welsh he would ease the problem of the Welshmen who could not express themselves as they wished in the strange language of the English. George was quite scornful of any mistake a Welshman might make when speaking English, yet he had a poor grasp of his native tongue himself, and made a terrible mess of any difficult expression.

After some weeks of practice he could say two Welsh words with confidence. He would greet an acquaintance with "*Shu mae*" when he wanted to know how he was faring, and often used the Welsh word signifying "finished"—"*Cwpla?*"

A middle-aged man who could speak no English came to work at that colliery. During the first shift he was instructed to clear coal in one stall, then go along to another place. He was keen to make a good impression, so he cleared the first lot quickly. When he wanted to go to the next place he asked two workmen, but neither knew what he wanted, because they could not speak Welsh. He was worried and wandered about the stall because he did not know where to go next or how to explain to these strangers. Then, crawling out from under the low height of the coal-face, came George. We were on the way down, and George was preparing in front. He stepped out to the better height of the stall road, then straightened his aching back and legs. He noticed the dismal stranger and greeted him:—

"*Shu mae?*"

The stranger was all attention at once, for here was a man that greeted him in a way he understood.

He beamed at George, who, seeing no coal in our way, exhausted his Welsh vocabulary by asking had he finished?—"*Cwpla?*" The stranger was convinced by that word. He stepped forward, grabbed George's arm, and rattled off such a dose of questions and complaints that George—although he failed to understand one word—was convinced that something serious was the matter. When he did not reply, the stranger started again—even more urgent and emphatic this time—and, in his eagerness, he pushed his face close to George's, while he shook his arm and prodded him on the chest to help the explanation. George was dumbfounded. At last he broke away and scrambled back to us; there to explain:—

"There's a queer old bloke up in the next stall, boys. Grabbed me tight, he did, and was blabbing a lot of nonsense like old boots. Best for you all to come with me and see if he's gone off his nut."

Fortunately Hutch could speak Welsh fluently. George has since given up his interest in that language.

In one way it was good that we worked so many hours, because the more we were in work the less we felt afraid of the risks. Almost every day our backs were soaking wet by the end of the first half-hour, yet we would not be able to have dry clothes or warm food for another sixteen hours and more.

Nearly always, while we were cutting, the top was squeezing downwards above our heads. The posts would be cut about five inches shorter than the coal-height, to allow for a wooden lid to be driven on top to tighten the post. It might be no more than ten minutes later when we wanted to stand these posts, but we very often found that the top had dropped so much that the post was too long to be placed in position without any lid at all. We could detect no sound of this squeezing, for the noise of the machine at work was so loud that we could hear nothing else, and even after we had been home some time our heads and ears still held the sound of those picks bashing into the coal.

Somehow we avoided serious injury, although the falling coal and clod, the tearing chain, the bursting cable, and the danger of that amount of electric current made every day's work full of risk. During one week I was completely buried on three occasions, but the nearness of the gob-walls and the height of the machine saved me from being squashed, and they got me out with only a few scratches each time.

Instinct warned me once, and I dived away at the same instant that more than sixty tons fell on the machine. I hurried to the gate-end switch to shut off the current, and I had gone so swiftly that the two men working behind—not ten yards away—thought I must be buried. I was returning from the switch when Hutch came rushing along, and he was so frightened that he did not recognise me.

"Help wanted in there, quick!" he gasped as he dashed by me, and my puzzled "Where?" was not answered, for he was too far away. I noticed a light in the stall where the machine had been cutting, so I went in there again.

Tommy had come in, and he was hurling stones—surprisingly big ones for a man to move—behind him and gasping to himself as he worked:—

"He's under there. Oh, God, let me be in time! Let me get him out alive."

Tommy was whimpering with fear—and he was not the sort to do that without cause.

I stepped up close and asked: "Who is it, then, Tommy?" because I thought someone else might have gone into that place after I had rushed away.

At the words he stopped lifting, then turned to look who had spoken. I have never seen a man so surprised. It was some seconds before he was convinced that I was really alive, and not under that fall. Hutch hurried back with some help, and we were bound to laugh at the mistake, but Tommy and the rest of us took a couple of days to get over the feeling.

These near escapes make us nervous for awhile, especially if we think what might have happened. For some time we see a danger in every stone, then become hardened again. I remember when a boy of seventeen was killed about twelve yards away from me. More than sixty boys were working there, but not one came to work on the next shift—all were afraid.

I can recall three more instances about that time

when I had near escapes. These happened when we were short-handed; for if we did not have a lot of cutting to do, they were in the habit of sending two of us home, and keeping the other two on to finish the cut. Tommy was working behind the machine that day, but he had gone to hunt for timber. I was driving slowly along, and getting behind the machine as often as I could so that I was able to stand a post or clear the small coal. We had not a great deal of cutting to do, but we did want to finish and go home. I had travelled along several yards by myself in this manner with the machine in slow gear while I was at work behind and in top gear when I got back into my place with the levers; then I came to a place where the gob-walls were too close against the machine for me to struggle past.

In the ordinary height there would be about six inches from the top of the cutter to the roof of the seam—not enough to allow of a man's body passing over. In this close area it meant going all the way down that stall road, along the heading, and up the next stall, if I intended getting behind the machine. This was much too long a journey; it would delay me too much. Then I got under a place where a small section of the roof had fallen, and so more height was left over the machine. I saw a chance to crawl over and get at the rear. I set the haulage control at half-speed and locked the other levers, then took the sledge in my hand and drew myself up on to the top of the coal-cutter. Even under this extra height there was nothing to spare above my back, while in front of the machine was an especially low part that was not more than four inches above the steel plates.

I slid along the warm steel plates, keeping well to

the right of the picks, which were whirling around so swiftly that it was beyond human sight to tell whether it was a pick or a section that was passing, and timed myself so that I would drop over the sprocket end and get ready to put a post up. I was square on top of the machine, with my back against the top and my stomach on the plates, when I felt something was holding me. I tried to slide forward, but could not. Another wild jerk did not free me; then I knew I was fast. I guessed—and rightly—that one of the nuts that was fastening the plate down had gone inside the button-hole of my waistcoat and was holding so that I could not go on or back. I could not raise myself enough to get free, and I could not turn back or reach the levers.

With every second that mass of steel was panting along, drawing itself along the steel rope, and with every jerk getting closer to that very low part where there would be no more than four inches clearance. I was quite fast, with no way of freeing myself, for the strong waistcoat resisted my efforts to tear it. Already I could feel the pressure on my feet increasing, and I knew that even at that slow speed I would be under that lower part in two minutes, and all my body would have to go through a space that was only four inches in height. I screamed again and again, and struggled, but failed to get loose. There was no sound or sign of help, and even if someone had come, I was doubtful if they would realise what was happening in time to save me; besides, it would have to be one of the few men that were familiar with the working of the machine and knew how to stop the engine.

I had released the sledge-hammer in my struggles, and it lay on the top plate just above the jib. I slackened back as much as I could, then dived forward in a last desperate effort. I was frantic with fear, for I could almost sense my bones being crushed under the top. My arms flailed out like a swimmer's, and the left one knocked the sledge-hammer over the side and on to the chain. The picks cut the thick handle apart from the sledge-head as easily as they would have sliced cheese. The sledge-head danced up from the picks, then dropped on them again, and this time it became locked between a pick and the solid coal. Crash! In that sudden locking the chain was forced to stop, and the strain was too much for the electric switches. They blew out on the gate-end switch. The roar and pulsing forward of the machine stopped on that instant—all was silent. I took some seconds to realise what had happened, and lay there with my feet fast and my body quivering. I was some time getting free—to tell the truth, I was trembling too much. I kept that sledge-head for some years after; it had the marks of the picks driven into it. Never again did I try to crawl over a machine in low height.

Sometimes the roof would fall and leave a hole upwards of fifteen to twenty feet over a wide area. They would clear a gully of about two foot alongside the coal, and we would have to cut past and post it up. We were cutting alongside one of these falls one afternoon, and Tommy was the only one with me. Again we were short of timber—the usual complaint—and no fireman came near to us, for fear we would insist on him sending help to carry some. I had noticed that Tommy had been away for awhile, then he came back and signalled me to stop. I did so, then he explained that he had found posts at some distance, and he had loaded them into a tram. He could not push this tram along the heading, so he wanted my help—I was the only one he could find.

Everything was silent when we went, and everything was silent when we returned. We were not more than ten minutes away, but in that time over a hundred tons of stone had fallen above the machine. It took the repairers, working at forced speed, nearly twelve hours to clear a narrow way to the front of the machine—where I would have been crouching.

Another afternoon I was guiding the machine along when I heard a thud behind me, and saw a cloud of dust go past. I looked round at once, but although I knew that George had been preparing posts there a few minutes earlier, I could then see no sign of his light. I knew something had fallen, and dreaded that it might have caught him. I switched off, then crawled up the coal-face shouting "George—oh?" I had no reply. Everything was as silent as it can be half a mile under the surface of the ground. About thirty yards along the face I found what had fallen. It was a slab of clod about six feet wide and more than twelve feet long, with a thickness of from eighteen to twenty inches.

George had been working under that piece, but now I could see no sign of him. What if he was underneath that stone? I called again, and neither George nor anyone else answered, and indeed, as far as I was aware, there was no one besides him nearer me than the stoker outside. I tried to look under the stone. The small coal prevented it going flat on the bottom. I tried to move it—in vain. I did not know which was wisest—whether to try a rescue myself or run to get help. The distance of the journey to get help decided me to do what I could myself. I began to scrape the small coal away from under the stone, and when I had cleared enough, I had to pack smaller stones underneath the big stone to prevent it dropping lower. What a rotten feeling it was—to expect that every handful of that small coal I scraped would show the body of my mate lying crushed, or that I might touch him with the next reach forward. After some minutes I had to be careful for my own safety, for, as I was lying on my side and stretching under, it was possible for me to get fast in my turn. I cleaned almost right underneath before I was assured he was not there. He returned with a load of timber soon after, and explained that he had forgotten to tell me he was going. My spoken opinion of him was in direct contrast to my feelings when I was hunting under the stone.

Every Saturday morning we went into the official cabin to give our time—and there was a row. All the week they would encourage and force us to work on and on, but when it came to paying—ah! then the band played.

We always had our time written out, and spoke as one man. The under-manager would not deny the time we had worked—Sunday work was double time then, and overtime paid as time and a half—although we often had twelve turns booked. He had only one reply:—

" They ain't willing for me to book more than ten shifts."

He could not deny that he had forced us to work, and that we might have been sacked if we had refused. It did not matter, ten shifts was the limit, and we rarely got anything more, unless it was half a shift extra on an exceptionally good week. We begged them—as we had often done before—to put another shift on, but they would not. After the heat had gone out of the debate he would close the book with a sigh of relief and say:—

"Well! That's that once again, boys. You'll be here as usual, seven o'clock Sunday morning, I s'pose."

He knew he was robbing us, and was decent enough to feel ashamed, and we knew that his statement was true—that he had to do it "because of the orders from headquarters". The policy of the mines is very often controlled by directors who live many miles away and rarely dirty their person by coming near to them.

One time, when Hutch's wife was very ill, we had been at work continuously for nearly twenty-four hours. He was worried about her, and was miserable. We suggested, as we had the cut nearly finished, that he should dodge out through the return airway after the fireman had passed, and get home in front of us. We would do his work between us, and cover him by some excuse if the fireman happened to return. Hutch got away all right, and it was still dark in the early morning when he was outside the level-mouth. One of the tippers was at work on the tip. Hutch did not want to be seen, so he crept past the cabins and got to the bottom of the tip as this man arrived at the top with a tram and horse. Looking down through the faint light, the tipper saw something moving—something that had the shape of a crawling man. He shouted—a very nervous hail, that was, of course, not returned. The second call, even more shaky than the first, fared no better; but by that time the tipper was sure that it was no living man that was moving about in the shadows. He thought of the many fatal accidents that had happened there recently, and gradually fancied that moving figure into the ghost of one of the dead men. He jumped into the tram, whipped at the horse, and galloped inside the level, not slackening the speed until he had an audience for his exciting tale. He would not work by himself on that tip afterwards, and was always looking fearfully across the mountain for that ghost;but he was never honoured by a return visit. We never told what we knew.

We had another ghost-scare soon after, and although some of the men thought there must be a connection, I knew the facts of the second scare, too. One of the colliers had come out late on several days, and decided to alter it. He brought to work with him the only timepiece he had. It was a small mantel clock, little bigger than a watch, but with a very loud tick. In his working-place he put this clock inside a large tin to shelter it from the stones and the wet drops from the roof. He went out in haste and forgot the clock. The tin made the already loud tick of the clock even more emphatic, and the hollowness of the underground workings emphasised the sound still further. It was not much wonder that the afternoon fireman was frightened by that strange and persistent tick-tock that he heard in a place that should have been silent. He retreated to the main parting, and brought a haulier on with him. This haulier not only supported him in his hearing, but defeated him in the race back when they were both convinced. No one went into that place after that during that afternoon shift. The workman had told his mate on the night shift to look out for this watch he had forgotten. When this mate arrived at work that night, he was told, with great emphasis, about those queer ticking sounds that had been heard in his working-place. He guessed the cause, but was a bit of a humorist, and refused to work in a haunted place unless the fireman stayed the night with him, for, as he argued:—

"It's the colliery company's works, this is. It ain't my fault as the place is haunted. If there's a ghost there it's sure to belong to the company, for they claims every blinkin' thing, and it's their look-out to keep their ghosts in the proper place."

The fireman would not stay the night, so the workman had a very easy shift. The clock had run down by the morning, and was taken home secretly.

I had enlisted under the Derby Scheme and been passed. I was sent to Cardiff to join up, and was then sent back home as indispensable. By that time I was getting quite fed up with machine roars and long hours, and felt quite annoyed at being returned. I remember the doctor's remarks about a man who had been rejected just in front of me. He was annoyed at that man being sent to him to waste his time.

"The fellow is dying," he said, "can't last many weeks."

That fellow has reared a large family since then, and seems little more likely to die than he did then.

CHAPTER EIGHT

JUST after the end of the War we had a wonderful stroke of luck—so unexpected that we could scarcely believe it to be anything but a dream: we got a house to ourselves. We had preference over a large number because the renter was a close friend of my wife's parents. I suspect that a deal of coaxing had to be done, because people there were prepared to go to any extremes in bribing or paying extra rent to get a house that became vacant; but, unbelievable as it was, we got it.

It was a small house—white-washed—with two small rooms down and two up. No pantry or scullery. The front door opened towards the main road, and the other door in that room opened direct into the kitchen. There was no passage, so that if the front door was opened at the same time as the back one, we were lucky if we did not have to buy a fresh lot of dishes to replace those blown off the dresser. Having no back entrance, the returning collier, and the load of coal when it came, had to be taken through that front room, of which my wife was so proud. The floor was below the level of the street, and water ran down against the front door. At the back it was too low for any drains to act, so we had to carry all the rain-water, while in the winter we had to have a plank over the water so that we could get from the back door to the garden. We came down on several mornings to find the water in the kitchen, and during one wet winter we had the water flowing in through the back, while we swept it out the front as fast as we could. On four different days it filled up in the room high enough to put the fire out. When I did not step into water in the mornings I stepped on to black beetles, for there were thousands there, and they remained there, although we tried everything to clear them. We could hear the papers rustling in the night while they moved over them, and the pockets of my working clothes, and often the boots, were favourite places with them.

Despite its many drawbacks, it was a house to ourselves, where we could talk without being overheard and where the girl could play as she wished. We counted it as almost a palace, and besides, there was the small garden.

What visions we saw in that garden! It was covered with ashes and adorned with many empty tins when we went there, but not for many days after. We cleared all refuse, dug it very deeply and manured it, then laid out paths and flower-beds. I know a deal about gardening, but was outdone by my wife, who is about the most expert woman gardener I have ever met. The only book she was interested in was a gardening catalogue; the only picture she could see was a well-arranged flower-garden. She planned it out on paper, and nothing that she planted ever failed to grow. She would tend and water them, even give them water in which an aspirin had been crushed, with the same tender care as if they were children. Every spare minute was spent in the garden. Within two years we had a show place with standard and climbing roses, asters, dahlias, chrysanthemums and any other flowers she could beg, exchange, or stint to buy.

Next to us was a large public-house. All day long the thumping of the pianos and the demands on the barmen sounded through the walls. Outside there was the rattle of the traffic and the daily dumping and rolling of beer-barrels. It was a hopeless place to sleep on the frequent occasions when I had to use the day to try to make up for the nights spent in work.

In the years after the War we waited for that better country and conditions we had been promised. They were slow in coming, and the "audacious demands" we had been invited to make when peace was concluded did not seem likely to be granted.

We were delighted when Bob Smillie, assisted by Frank Hodges, exposed the coal-owners' case so completely at the Commission in 1919, and Justice Sankey's findings in our favour convinced us that at last we would have some measure of justice. I remember the pleasure of coming out at finishing time on the day after the seven hours had become law and finding that we had an extra hour of rest.

It seemed that even with the low cutting price of their coal, the company were not satisfied. Another small colliery close to ours had forced their workmen to loàn—that was how they described it—the owners half a crown a week from each man's wages until better times came.

If the men earned two pound ten for a week's work, they were compelled to hand back half a crown out of this wage to the company. Our company fancied a few spare half-crowns themselves, and demanded the same off us, but we objected strongly, and did not give anything away until we had been idle for three days. Even then we would not agree to returning anything from our wages or cutting the list; instead, the colliers agreed to fill a tram of coal a week for nothing. Twenty-six stalls gave the owners trams of coal free every week.

Then the company decided that the cutter men should make some sacrifices, and insisted that we work all our overtime without any extra above the day rate, and that only time and a half should be paid for Sunday work. We refused, and stormed a general meeting with our complaints after knocking off work early. The committee had plenty of complaints of their own, but they promised to consider ours; but we did not get any more than the company's rate afterwards.

I have often heard the instance of the anthracite, or stone-coal, miners in this district quoted when there has been a question of the miners giving something away to help in bad times. Before the War the anthracite mines were not prosperous, and the men agreed to a temporary loan of five per cent. in the wages—as a reduction for the bad times—to help the companies.

After the War the anthracite mines boomed, and the coal fetched almost double the price of steam coal—yet the men have never been given back their old rate. Miners quote this instance to prove that what they give away they never get back—and I have seen it that way.

All the expenses that go in the running of a colliery are added together and called the cost of production. Into that cost of production must go the wages of those relatives of officials who are kept in well-paid jobs when the colliery would have gone on just as well—or better—if they had stayed in bed. But cost of production is a convenient and most elastic item. Even with an inflated cost of production there must be a lot of money being made out of coal by people who do not risk their bodies to get it. I have a friend who is a delegate for his Lodge to the conferences. At one conference which was held at Margate he saw the same coal—named and specified—that was produced at the colliery where he worked listed in one of the coal-agents' windows at two pounds five a ton. A week before that he had found out that the ascertainment board had stated that the total cost of production of this coal to pithead was seventeen shillings a ton. He was confident that seventeen shillings was a high estimate, yet between that coal being placed in the truck and reaching the consumer, someone was sending its price up nearly treble.

Many colliery companies act as their own sales agents, their own shipping agents, their own timber importers, and probably have large interests in the railway companies. What does it matter to them if they sell their own coal to themselves at a loss, as long as they get a good price for it when they sell it again as agents, then get freightage for shipping it, and make a profit from themselves by selling themselves timber. The more profit they make on the timber, and the more the sales agent makes in commission, the more of a loss the books of the colliery will show.

I wonder why our leaders always hold the conferences at pleasant places like Blackpool and Margate? I have heard men ask one another that question many times. Why not hold them in the Rhondda Valley, or at Landore or LLansamlet, or on the Tyneside?

Every mile away from these places makes the memory less dreadful, and every year our leaders spend away from the very heart of the industry makes them feel more contented with the conditions of those they represent. Thirst does not feel so terrible when one has found a running stream, and who fears hunger after a good meal, and with the knowledge that food will always be available in future?

The place to learn the truth about the mining industry is where the sick and crippled miners struggle along the rough roadways in the hope of finding a little coal to warm their bodies, and where the haggard mothers watch their children go to school and hope they will grow up quick, but not get very hungry; or that they will not play too much, because they cannot say from where the next pair of " daps " is coming.

I think it was in 1920 that they started to try conveyors in the coal-face. They used the trough type—in which little cradles under the troughs held wheels which rocked the conveyor in a smooth motion one way and a jerk back, so that the coal was worked farther along the troughs. It was necessary to have a slight slope towards the tram, and to have a roadway so arranged that fresh trams were continually moving under the trough end, for it did not take longer than two minutes to fill a tram.

Conveyors save the cost of ripping and laying roads: the stall-roads are done away with, and the coal is conveyed down the coal-face, often a distance of a hundred and eighty yards. Without the frequent stall-roads there is not such a good chance to escape from a fall, for under the conveyor method a man would have to crawl under a low roof that was moving above him before he could stand upright and run for safety. That crawl might be for a hundred yards. Timber had to be dragged or pushed along the low face, for there were no loading-places every few yards, like when the stall-roads were in use.

The layer of small coal left by the cutter—about three inches deep—made the height under the top even less than usual. This small coal was thrown over the conveyor into the gob, and a lot of the large coal was thrown there too, at the times when they were in a hurry to get the cut out of our way and trams were short. I have seen men working busily to throw what must have been more than ten tons of large coal into the gobs, where it would stay for ever.

Sometimes the rubbish-trams would be half full of coal and half muck, but all of it would be shot over the rubbish-tip outside the colliery mouth, to be buried completely by the tippings of the next shift. Yet if anyone wanted coal badly and took a few small lumps from one of these tips before the rubbish covered it, he was summoned and taken before a magistrate, who would lecture him about the crime of stealing, and fine him heavily. The men felt bitter about this matter, and I could never count a man a thief because he took what the others threw away.

During the early days of the conveyor working the under-manager stood alongside the tip end of the troughs and timed the trams as they were filled, while the fireman crawled up and down the coal-face hurrying the workers up and making sure the troughs were kept full. Often these troughs were set so close to the seam that the men had to crouch tight against the coal to avoid being hit by the jerk of the conveyor, and if some coal happened to fall where they were, the only thing they could hope for was that the large lumps were few, else they might be crushed between the coal and the conveyor. It was necessary to "shift the conveyor" every day. This meant moving it across about four feet nine into the track where the last cut but one of the machine had been taken. It was sometimes a very dangerous job to unscrew and move those long troughs across in the low height. They were made of steel, so could not be bent round the timber. Posts had been placed every two feet after the machine, then enough of them had to be knocked out to allow of the ten-feet-long troughs being moved across. These posts were replaced, but the moving disturbed the weak top, and the men who "shifted" had to do a deal of hurried rushing back to safety and cautious crawling back to their work.

There was a lot of noise when the conveyor was working, too, and sometimes we were cutting right alongside them at the same time. The roar of the machine, the clang and rattle of the conveyor, the noise of more than thirty shovels being used at the same time, the sound of the trams being moved, and the shouts of the hauliers, made up a bedlam that was not very soothing when one knew that one should have silence to listen if the top above was moving. The more mature miners would not risk themselves on this work because of the added danger through the noise. Then there was the dust from so many fillers, and the danger that the cable might be pierced by the picks of one of the colliers, or that someone might be caught by the jump of one of the troughs.

It was known that the top was too weak for conveyors to have a real chance, yet they kept on forcing the pace for weeks, and were soon trying to find out what was the lowest possible price they could pay for this conveyor filling. Before a new price had been arranged, we went in one afternoon to cut, and met the coal-filling shift coming out. They told us everything was clear, and we could cut through. When George went first to drag the cable along the run, he called on us to get back quickly, because " the damned mountain have fell in here ". That was an exaggeration, however, for it was only a few thousand tons of it. Less than ten minutes before it fell thirty men had been too busily engaged underneath it to hear any sound of its movement. We had several of the troughs out, but left some there for ever. That was the last time they tried the conveyor up the Rise side. There was some dispute about the royalty that was to be paid on the coal from that area, so they left that district alone for awhile, and we went down the Deeps as colliers—very gladly. We were able to finish work at the end of one shift, and George could hardly get used to that. He said " things are getting slack at this colliery now. We're only working one shift at a time."

We had leisure to walk about together and enjoy the pleasures of the village—which were few. Some of the men took the doormat from its place and sat on it instead of washing. They would sit or lie on it and talk to one another while their wives did their own bit of gossiping or waited their turn for the bakestone. Why they called it a bakestone, I could never make out, for it was made of iron. It was flat, pretty heavy, and had a strong handle. They made round Welsh cakes with it, and they were nice. I used to enjoy the smell and taste of these, and loved to watch the flames licking up around the bakestone while the flour-paste gradually cooked. There seemed to be a sort of communal ownership of this bakestone, and there were mostly a couple of women waiting their turn to cover the handle with a rag and take it home for their own use. They sampled one another's cooking while they waited.

The weekly visits of the cockle-women were the cause of a deal of friendly talk. They were quite characters, these strong women, with their calls of " Cockles *hethi*? Nice cockles to-day? " They could balance a wooden tub on their head and carry a full basket on their arm. They spoke of the dangers of cockle-getting when they had to ride out into the sea on donkeys, and the risk of sudden storms. I had a lot of trouble with donkeys when I was a boy, but I did feel sorry for any donkey that had to carry some of these women. They brought laver-bread, or *bara lawe*, in their baskets as well as cockles. This was black, slimy, stuff that everyone except me seemed to enjoy.

We had a picture-house in the village—only one. As there was no opposition, the proprietor was not particular what sort of pictures he showed to us, or how he showed them. I think he got them after everyone else had used them and taken a bit from them. " Pathé's Gazette " would snap in the middle of a procession, and when the picture started again we would be in the second part of a romantic story that, after a few more snaps and delays, would finish up in a wild ride of cowboys and Indians shooting things up in real Western fashion. But they were enjoyed.

One stout woman assured me in the middle of one of these pictures:—

" Them be devils, them be "—she referred to some Indians who were doing a little friendly scalping, and hoped to comfort me by adding—"but don't you worry, boy *bach*; ours will be here in a minute."

Ours were the cowboys.

Another man I knew saw the heroine foiling the villain by hitting him with a chair. Soon after he attacked her again, and her slowness was too much for this onlooker. He jumped up and shouted:—

"Why don't you pick up another b—— chair?"

Pearl White was a sure favourite. We had her in serial after serial, and spent anxious weeks waiting for the next instalment, to see if she escaped. We sang songs about her, and if any other picture failed to arrive, it was quite all right to put Pearl back on again.

Sometimes we had flesh-and-blood companies there, and I always went to see them, whether they were dramatic or operatic. Real acting always pleased me better than the shadows. I felt these companies never had a fair chance, for they were not allowed the best nights, and so had little hope of making their expenses.

The people with whom they lodged often had to be short of payment; sometimes they had to give food to save the actors from starving. How could anyone be expected to give a good performance under these conditions? When their showing did not please him—and he had strange ideas—the proprietor would sometimes stop them in the middle and order the operator to "put on Pearl".

I remember one company that tried to act "Bo-peep". I heard a man ask the proprietor what sort of a show they were giving. He had just been inside, and seemed annoyed as he answered:—

"Bo-peep be b—— I'll Bo-peep 'em!' They can't see, let alone peep. Hey, you"—he waved to the picture operator—"just shove on Pearl—I'll learn 'em what acting is."

So Pearl went on again, and the flesh-and-blood company went off, and that is about the history of the theatre in our village.

CHAPTER NINE

SOMEHOW we could not believe that the lock-out in 1921 would happen, and when it did we felt stunned. We were not then hardened to strikes and stoppages. We watched the company stacking its coal ready and making all preparations, yet hardly realised that things would be so serious.

We had some money in the bank, and my wife had caught the craze for shop-keeping, so I had made a counter and shelves, and our front room was a shop—it was well stocked too. We had reared two bacon pigs, and we were selling the bacon, and had pens and poultry out in the garden. We gave the things from the shop until they were all gone. We had very little cash for them at that time, or afterwards. The debtors, too, were not all people who had been thrown idle by the dispute and had no money, for one assistant manager at the Labour Exchange went away owing us more than a pound, and they would not give us his new address.

We paid the wholesaler's bills, used the rest of our money, then did as almost all the workers here were compelled to do—went to get parish relief. All the long shifts and the weeks without proper rest, all the dangers I had passed through and the dust I had swallowed, had gained me this much: that I had to go and beg for parish relief. My dreams of a little small-holding in the country, for which I had gone without pleasure, without a drink or a smoke, were all useless.

We had to go early in the morning for this relief, because there were hundreds in the queue outside the chapel where they gave out the relief-slips. I wonder would it cost them very much effort to be only decently civil to the ones who apply for relief? Or do they purposely pick men who have no sympathy or patience? I did not hear the officers speak one decent word during that period when we had to go there weekly—they seemed to hate the applicants. It seemed that we were asking them to pay out their own money. Yet the relief we had was supposed only to be for the women and children, and the men were allowed nothing. Even that little was given on loan, and we had to sign our agreement to repay it when the stoppage was over. Surely there was no need to snarl over relief given under these conditions to men who had a long record of hard work?

After the event of the week—fetching the parish ticket—had been endured we had very little to do except sit on our heels and discuss the stoppage and the prospect of re-starting our work. We walked up and down the roads until the grit was worknig through our soles, then we made soles and heels out of old motor-car tyres, and found them very serviceable. We wore out packs of old cards; we played shove-ha'penny with buttons; and sat for hours playing tippet with a stone. I read continually, gradually abandoning Western yarns and finding pleasure in serious reading.

I was given a pile of old books one day by a shopkeeper. He told me that he had not bothered to read them, but I went through every one carefully, for I hoped that there might be one good one amongst them. There was—George Borrow's *Wild Wales*.

I did nothing else until I had finished reading it; then started the second time through. I took it with me up the mountain, and delighted in his account of the Welsh poets and his translations of their works. I took it with me while I followed in his footsteps along this valley, enraptured by the mountains as he had been; concerned over the bridges; disgusted at the noise of the train; and finally climbing up the long slope and seeing the light in the sky—my old inspiration—guiding me to the journey's end.

We hunted for wood on the mountains, and opened levels so that we could have coal for our use. This was not permitted, but we took the risk. Very grave risks some of them were, too, for a little coal. Most of this coal was taken from the bottom of an old pit, which we descended by sliding down an old rope that might have snapped at any moment, and did not seem strong enough to wind one small bucket of coal. We had to hope for the best of luck, for we had very little timber to put the roof safe, while the sides of the shaft threatened to collapse over us every minute.

There were several accidents, during that stoppage, in the hunt for coal. One man had his legs broken, and we had to carry him across the mountain on a hastily made stretcher. Another young man had his back broken. He was not entitled to any compensation, so he has had to be kept by his parents ever since.

We cut one another's hair, but kept our caps on for a while afterwards, unless we could persuade someone who was more expert to repair the damage.

We had one solace during the stoppage: the weather was kind to us, and the sun shone day after day. I went across a stretch of waste ground behind our house one morning and saw two grown men amusing themselves by knocking a little ball about with a home-made bat. I joined in, and soon we were half a dozen running about after that old tennis-ball. That afternoon the son of the publican offered to sell us a complete set of wickets and a decent bat for five shillings—he had no liking for cricket. We debated this offer for awhile, then agreed to start a club—myself as secretary. We accepted the wickets when he agreed to wait until work started for us to pay him. Another man who had been a cricketer and still had his bat was made a life member when he offered to loan this bat. We got more members, and by stinting and collecting amongst the shopkeepers and others we kept up the payments of two shillings a week to pay for pads and balls.

Then we meant to make it a first-class ground. It was a fine display of communal effort: about thirty of us working from morning to night without a grumble carrying stones away, filling up holes, cutting and carrying and laying turf. We appointed a man as groundsman because he claimed to have seen the Oval. He went round with a pail of whitewash and marked the boundary of the outfield. That night—it was moonlight—the soft thud of an axe sounded into the early morning, and next day six small trees that might have been in the way of a good hit, and were inside the whitewash marks, had completely vanished.

A farmer loaned us his roller, and we rolled and trimmed, practised and planned, all day and every day. We could not play on Sundays, so we held general meetings that lasted most of the day, and were held near the cricket pitch so that we could watch the grass growing and stop children and cats walking over it.

After we had started we had matches every day, and the village came to see them. Surely never were a few shillings better spent. The married men among us who had small babies used to bring the babies there while their wives did the housework. If these men were players, and it came their turn to bat, another member of the team would try to pacify the baby while the father did his bit of slogging. Then, when we fielded, the spectators would mind the babies as part of the admission fee.

We crossed over the mountain for one of the matches. We walked, and carried our cricketing gear. More than a hundred and fifty of our supporters followed us the six miles there and six miles back. They saw us win a grand game, watched by —I should imagine—every inhabitant, as well as the four policemen in that other village. We had a concert afterwards, and it was nearly midnight when they let us depart. Six miles back over the moonlit mountain, and the game being played over and over again, while outside my house the secretary and captain of another club waited for us to promise that we would visit them on the morrow.

I will finish with the history of our cricket team. The second year we had a first and second team, with forty playing members who had whites. The first eleven won twenty-five out of thirty club matches. The third year was even better, and we bought a small shed and a large tent. Towards the end of that season there came a demand for us to pay a rent which we thought much too dear. We were prevented from playing there, and it went to look rough again. During the winter the banks were broken and the water washed over the ground.

Once again it became covered with stones. So now all the cricket we get is what we read about.

When the chill of the autumn of 1921 was making us notice how worn our boots and clothes were, there came an end to that stoppage. The miners had a big reduction in pay, and debts owing all around them. The colliery companies saw that they had their payments, and stopped them on the pay docket—a week's arrears of rent or electric light every fortnight.

They had decided to insist on a reduction of the royalty paid for getting coal cut at our colliery, and as it was not granted, they did not open for a while after the general resumption of work. I was told that this royalty meant nearly a shilling a ton to the landowner for every ton that was mined. I do not know if this included the wayleave that is charged for taking coal over land. As this landowner told them that the royalty had been the same per ton under his father, and would be the same under him, and the colliery company said they would not re-start unless there was a reduction, we appeared to be in for a dismal winter.

When all the miners were idle we felt a sort of companionship with one another. We were all fighting for a principle, and would begin work when it was decided. Then the others re-started work, and we were idle. We felt the bitterness of being unwanted. The hooters were blowing for others, not for us. They would have a pay on the week-end, but we would have none. I used to lie awake in the morning and listen, with despair in my heart, to the rattle of heavy boots on the streets when the other men went to work.

I went about every day, up and down the valley, seeing if work was to be had, but there were hundreds more on the same journey. There was a waiting crowd every day outside the colliery offices. It is a rotten feeling for a man who is working to come outside and see that a crowd of men are waiting for work. It warns him that the masters can treat him as they wish, for he dare not insist on his rights when there are so many waiting for a chance to start. Some of the officials exploit this fear to the limit.

What a worm a man feels when he must put a beseeching note into his voice when he asks for a job! It is not that he is afraid of the man or the work, but he is afraid of the refusal: of being told that his skill is not needed, that his strength and brain are useless, and that he must go home again and confess that he is not wanted.

The mine officials—in many cases—have been born and reared in the area in which they are working. In that time they have made friends and enemies. What a chance they have to show their authority and spite to anyone who has offended them in argument, or religion, or even in loving in the past, and is forced to ask them for work!

At one time the man was compelled to get the signature of the manager to show that he had applied there for work. Apart from the amount of extra work this gave the managers, it meant that the men had to wait about in the rain and dust for hours to see them, and if they wished they could avoid signing. Is it necessary that men should stand in crowds while an official picks one or two out as he would a sheep at the market? Naturally he would not pick the weakest—or the most intelligent-looking.

I know one manager who made it a rule never to re-start any man who had once left his employment under that firm, no matter why he had left it. He would look at a group of men and comment loudly:—

"Ah, yes. I could start a couple of you. Let me see, have you ever worked here before? You have? I thought your face was familiar. Ah. Never mind. Nothing for you. And you: have you ever worked here before? Never? Very well, you can start to-morrow."

He was a very forgiving type of man, he was! But the men had some consolation years after, for he was sacked himself, and failed to get fresh work.

I knew one man who walked to the same colliery every day for over twelve months and waited for the manager to come out. Every afternoon he would step eagerly forward, then the manager would look at him and shake his head. No words were passed between them, but he was always there, and the refusal was always waiting.

I met many men who had not worked for years, who did not expect a manager to say yes. They were losing their skill and fitness, while they knew that the pace of mining work was increasing every year and the kind of skill necessary was altering with each fresh invention. The older men would never settle down to work under conditions that left their safety to chance, for they had seen things happen that the youngsters had not; and they wanted to trim and work the coal in a neat fashion that was not suited to the times when it was becoming the custom to "drag the coal out by the hair of its head".

I used to feel my anger boiling when I heard a tender-handed Labour Exchange "snapper" ask a man old enough to be his father:—

"Why don't you look for mining work at a distance from here?"

Yes! Go away from the people he knew and counted as friends, who would smoke their pipe of tobacco with him while they talked of the better times that they had enjoyed together. Go to some strange area where he would feel himself a stranger and an interloper; to have his head and hands battered by coal and top that behaved in a way that was strange to all his experience. Go to a distant colliery with the knowledge that he was past his best, and that the conditions in this new place must be really hellish, or there would be no talk of vacancies. Go to be a stranger in the house, in the work, and in the street. Men get to think that they have some sort of interest or stake in a district where they have lived and worked for a long time. They feel that way, although it is mostly only a pleasant illusion.

I used to think that it would be a good thing for all if we could change our work around every few years. I am sure we should all learn more tolerance, and sympathy for one another. What a help it would be if the Labour Exchange managers had all suffered a long spell of unemployment themselves, and if the fear of it was not completely removed from their minds!

About a month before Christmas, when we had begun to talk about the poor prospect of having any sort of a festive time, I got work. It was either good fortune or cheek that got me in where other men had tried much longer and failed. We had walked four miles to a colliery that morning, but although we were early, there were quite two hundred men waiting, so we went on to the next works. About this time the companies had to employ police to keep the men who were looking for work off the pit-heads, because they were so many that they hindered the work. The men were supposed to see the manager, but were prevented by the police, so that they were between the Labour Exchange Committee and the Police Court, and could take their choice under which they were penalised.

While we were waiting in the queue at this second colliery I heard a man whisper to his mate that he had heard they were going to re-start the four-feet seam at the Perry Colliery. I knew the situation of this place, and persuaded my mate—a North Welshman was with me that day—to walk the extra two miles to that colliery.

Things were quite normal at the Perry Drift. The usual groups of men were waiting to take the place of the men who might be sacked or injured; or were waiting for that prosperity that some of our national papers are for ever discovering to appear. While my mate kept the place in the queue, I wandered around talking to the men and admiring the engines. It was taken for granted that anyone who was about the yards must be looking for work, and the ones who were working were very sympathetic.

One tried to encourage me by saying that a lot of men finished there every week. I knew that no man would give up his job in that part if it was possible for a human being to stick it, and began to puzzle whether it was wise to ask for work, because if a man starts at a job, no matter how bad that work may turn out, and fails to keep it up, he will have a lot of trouble in getting back on the dole. In any case it is likely that he would be punished with eight weeks' suspension—because an employer tried to take advantage of his eagerness to work. We had been drawing the dole from the time when the lockout finished and our working-place did not start.

I stepped out of one engine-house into a narrow gully that led to a path across the open mountain. I turned from looking at that mountain, then started to walk back between the row of engine-houses. At that minute a man came quickly along this gully. He was slinking along as if he had been a prisoner escaping from the police, but it was not wide enough for him to get past me, so he stopped, very reluctantly. I recognised him by a description I had been given. He could not pass unless I flattened myself against the brick wall, and I did not intend to do that for awhile. I stood squarely in his way, and asked if there was any chance of me having a job there. He looked at me with a whimsical smile on his face. I always found him quite ready to take a joke or see humour in anything.

" Collier, are you? " he inquired. " Used to the coal, eh? "

I said I was and where I had worked.

" I could start one or two," he agreed, " but I couldn't go and pick them from that crowd. What's the name? I'll write a starting-slip for you."

" What about my mate? " I asked. " He's about here waiting for me."

" Yes, of course." He was agreeable to anything as long as he was allowed to pass. " Is he used to the coal? He is? Very well, he can start with you."

" Well, good day now." After he had signed the slips and I had squeezed on one side, he was as cheerful as a bird. When he had walked a few yards he suddenly wheeled around and said, " I say. Not a word to those fellows out the front, mind. Or they'll be after me like a pack of hounds."

My mate was known, for that day, as Will Merthyr, although his proper name was Will Williams. He was ready to obey my signal to come out of the queue, although he did not know my news. There were more than twenty men waiting outside the lower office when we arrived to sign on. Most of these men were waiting to be paid money that they claimed was due to them, and we waited a short while listening to their complaints before we went inside. We learned that the yard-seam was a rotten place to work, and that the nine-feet and six-feet were little better. The agreed opinion of the men there was that no man could earn enough for bread in those seams, but one or two did say that if they could be made to pay the old price list on the four-feet when it got going properly, there might be a little cheese to go with the bread. Soon after we had heard this I saw a chance to get inside to one of the clerks, and Will followed close behind.

I handed the slip to this clerk. He seemed surprised and asked:—

" Had a start, then? "

I confirmed this, and he then inquired:—

" Did he say what seam? "

He had not done so, but I did some rapid thinking, then decided that the four-feet was the best chance.

" I'm not sure "—I hesitated as I said it—" but he said something about starting in the four-feet."

" In the four-feet? " He was doubtful. " Well, they are reopening it, I know; but I didn't think they were quite ready. Anyway, he ought to know."

So we were booked for the four-feet, and that piece of strategy paid us very well. After we had signed and started to walk home, we began to consider the difficulties of getting to work. We had over eight miles each way to travel, and there was no suitable train so early in the morning. I had a cycle, so I agreed to work on the day shift and ride to work. My mate could not cycle, so he agreed to work afternoons, and had to go from his house in time to catch the quarter-to-one train, so that he could be in work by three, as the next train was too late.

Three parts of the men in our village were idle, and our good fortune was almost as much debated as if we had won the football sweep. It was pathetic to see how keen some of the men were to get a start at the same place. There was quite a pilgrimage from the village up to that colliery for days afterwards. Men wanted to know if we thought there was a chance for more to start, whether we wanted to buy some tools cheap, or if we would try to put a word in for them as soon as we saw a chance. Every day some of them stopped me on the way home to ask how we were getting on and to question if any more stalls were being opened.

I had to get up about half-past four in the morning and be on my way before half-past five. The darkness and the wind and rain were all hindrances. I had to ride about seven miles, leave my cycle at a house, then walk nearly a mile and a half up the mountain; after that I walked some distance inside the colliery and started the real work.

My experience in holing seams came in useful here, for most of the men already working there were used to the easier working coal of the big seams, or else they were colliers who had always worked in the shooting-seams, and were accustomed to using powder for loosening the coal.

Right from the first week we filled a large amount of coal, and earned above the minimum wage, but in some of the other seams the men could not earn thirty shillings a week. I found things more comfortable here than with the machine-mining. I found that the officials were very reasonable and decent. The manager had lived at Treclewyd, and was fond of coming to my working-place to talk about the seams we worked there; but prize-fighting—as far as it could be done with the mouth—was his favourite subject. He would get very excited over it, too, and would demonstrate his favourite mode of attack by shadow boxing round the tram while I was an admiring, but distant, watcher. He was not very big, but he fancied he could knock all the heavy-weights out. I noticed, however, that when a workman no bigger than he was attacking him during a dispute about money, he hurried into the office and sent for the police. So it seemed there was no substance in the shadow of the manager's left leads, and his knocking-out was only done in theory.

Had he shown the same strategy in the ring as he did in avoiding payment, he must have won many fights. If a man went to him to complain that he had not been paid for a certain job, or that he was money behind, the manager would be disgusted at the treatment the man had been given. He would assure him that he had not known about it, and that he would not allow that sort of thing—oh no. The next week the man would still not be paid, and when he saw the manager again that official would be astounded—apparently.

"Not put it in again? What's the matter with the fools? No earthly reason in that sort of thing. When a man has done the work, he ought to be paid. But don't you worry, my man, you leave it to me"—that was his favourite assurance—"and I'll put a stop to tricks like that."

Many of the men believed him and waited, in vain; then they would blame the under-officials for doing what they were compelled to do.

There was a lot of shot-firing done in that seam, and it was handled very slackly. Detonators were to be had by the dozen; anyone who wished could take a handful. I found a bundle of more than a dozen in the middle of the roadway inside one morning. I was the first in, and some horses were next behind me.

Boys used to fetch the firing-battery, and use it. Things tightened up after a while, then they employed shotsmen and, strangely enough, it was after that change that most of the accidents with shots happened.

There were no electric lamps in use at that colliery then: every man used the same type of old-fashioned oil lamp. These lamps are difficult to use, for at the first shake the light goes out—probably when you need it badly. When the seam is warm or the air poor, the light will jump up in the glass, and that glass is blackened so that you have very little light. If the lamp slips just the least bit from the upright position, the flame dies out. There are good reasons for the lamp being so constructed, but a man has his work to complete, and it is awkward when he has a lot of work to do but has to borrow another man's lamp to find his way back a long way to where the re-lighter is placed; and it was a contract to re-light some of those lamps.

After I had worked there for about three months, I was asked to drive the main deep heading. This place was wet, and very rough, for awhile, but afterwards it improved and became the most profitable district, for owners and men, that had ever been worked there.

One day I had bored a top hole ready in the roof rippings so that our tram could be brought closer. The shotsman—he was not a careful one—arrived to fire this hole. I had pushed five cartridges of dynamite down this four-feet hole, had placed a cap in the sixth cartridge, and had pushed that against the others and had rammed clay tightly against the lot, only leaving the ends of the two wires that fit into the cap exposed for connecting to the cable. The shotsman is supposed to do—or personally supervise—this charging and ramming, but usually he has so many other men waiting for him that he hands the caps around and expects everything to be ready for him when he comes.

This shotsman slung one end of the cable down to me, then, taking the battery and the other coils of cable with him, he went inside a stall-road that was about twenty-five yards up from the face of the deep. It was his duty to connect the end of the cable wires to the cap wire, but this again is often done by the collier, so that time can be saved for both.

A new man had started to work in the skip just behind me that day. He was skimming the coal off for a thickness of three yards or so, then he would blow his road in and take his tram out of the way of my straight heading. The shotsman could not have been aware that this man was working there, and the stranger, on being warned by me that a shot was going off, took his lamp and walked up the heading to shelter in the road above where the shotsman was busy arguing with the stallman about the election that was coming. Powder and elections are both useful if not used together, but they nearly had a disastrous effect that day.

I dragged the cable near the cap-wires and connected one side of the cable to the red wire. Somehow, at that moment I felt a doubt as to whether I had placed my tools far enough away from the flying stones, so I left the other wire hanging, while I moved the mandrils and shovel farther into safety. I do not remember ever before leaving a wire loose at the time of firing, but this time my extreme carefulness saved me from being blown to bits and a shotsman having something awkward to explain—and forget.

Quite coolly I turned back, connected the second wire, and hurried to where the shotsman was waiting out of direct line of the shot. He asked me rather impatiently whether I was sure that I had made the connections properly.

"Yes, of course," I answered. "Why?"

"You can't have," he insisted, "because I've tried to fire the damned thing three times, and it won't go."

"Tried to fire?" I was astounded. "Why, I've only this minute walked away from it. Try it again."

He did try it again; the shot went off all right, and the top was ripped down. While I had been alongside the hole and putting the tools safe, he had been turning the key time after time in a vain effort to fire that charge. A second or two before I had made the final connection, and when that heavy charge of dynamite was right in front of my face, he had been trying desperately to send that complete circle of current round to fire the powder. Bare seconds had saved me, for if I had fastened the two wires at first, I would have received the full force of the powder.

He told me later that he watched the other lamp—carried by the stranger—go by up the heading, and thought it must be me, so he tried to fire. I went on with my work until about an hour before finishing time, then I had a sudden and very vivid realisation of what might have been the result. I went chilled all over, and felt sick at the thought of my body being battered by that powder. I could not do any more work that day, and felt really ill all the way home. I dreamed of that shot that night. I screamed in my sleep, and jumped right out of the bed in my fright.

I suppose I ought to have reported the affair, but the shotsman had a wife and children. He lost some days from work under the pretence of an illness, which was really fright, and was more careful for quite a while afterwards.

I had several near escapes from being blown up there, and in most cases it was due to the number of shots everyone was having fired and the rush to get them going. The one I recall most vividly was not caused that way, however. My mate on the afternoon shift had a hole that would not fire. The detonator must have been defective—in mining language it was a " flat " hole. It was at the end of his shift, so he had to leave it, but he called at the overman's house to warn him, and leave a message for me, about this unexploded hole. This overman had given me permission to leave my bicycle at his house because it was near to the incline. I cycled up early next morning, left the cycle as usual in the shed at the back, near which the overman was having a wash. It was about a quarter to six in the morning. The overman was always very decent to me, and he insisted that I should go into the kitchen and wait until he was ready to go up along with me. He made me drink a cup of tea, then, after filling his tobacco tin and putting the time-book in his pocket, he walked up to the colliery with me. We talked all the way, yet he did not remember the message from my mate. There was quite a lot of coal ready for me, so I decided I would go on holing in a corner which we had both agreed to work on, and which had been stiff.

I had a sharp mandril and I was fresh. After a while the coal began to get slacker, and the boy who worked with me—about sixteen years old—came up close and started to turn the free coal back while I cut more out with the mandril.

Suddenly I heard a rattle of wires, and the point of the mandril went into something soft and stuck there. I was puzzled, so I left the mandril idle and asked the boy if he had heard a strange rattle. He reached his lamp up close, and we looked carefully. We found some detonator wires in the coal, but were not surprised, because we often came across spent wires. These, however, would not come free when we pulled, and I discovered that they were still connected to a live cap which was in the cartridge of powder. My mandril point had gone right into this powder, but had missed the detonator by a fraction. The least jarring of that detonator-cap would have exploded it, and that would have fired the powder. Of course, we had not had any warning, and we sat back and started to express our opinion about careless shot-firers. Then we heard the sound of a running man coming through the darkness. It was the overman, who had only just then remembered. I noticed how the sweat was running down his face and the steam rising from his hair, while his limbs were twitching in fright. An eighth-of-an-inch deflection in the point of that driven mandril would have meant two deaths, the loss of his position, and probably a manslaughter charge against him—and the boy who was working with me was his youngest, and favourite, son.

Soon after this I was near to where another shot went off and killed a tall man, while the charge went right over the head of a very short man and missed him altogether. Another man with whom I had become close friends went to fire a shot, and took every possible precaution to be out of range. A stone rebounded from the side so that it passed round a right angle turn, where it again rebounded from the side of the roadway, and went deep enough into this man's stomach to cause his death after a few days.

The boys found the oil-lamps very convenient on a Saturday when they had a cricket or football match on. I used to instruct my boy in the mysteries of the late cut and the off drive, while he threw stones up to me and I did my best to be a graceful batsman when handling a shovel by a faint light. It was remarkable how many of the boys' lamps would fall down and go out in the dark about twelve o'clock on a Saturday. Strange, too, that it was not possible to re-light those lamps without going all the way to the lamp-room outside. Then when they were outside and found that there was no more than an hour to go before the end of the shift—well, of course, it was just as well to trot home and get into the cricketing togs. My boy had two severe attacks of pain in the stomach on successive Saturdays, but recovered in both cases, so that he was ready in his whites when I came down for the cycle. I did not mind in the least, for I liked to see him enjoying himself, and an extra hour in the sun was better than that time spent in the dust inside the workings.

I always liked to have a boy working with me,

and tried to treat them well. I was interested in their sayings and their reactions to the dangers of their new work. One thing I always feared, and that was the possibility of a boy being injured or killed while in my charge. Somehow, the boys seem very apt to get into danger, and the more they are warned the more frequent, it seems, they get into that forbidden place.

I had two men working on my left hand for awhile whom I found very interesting. One was a keen photographer and had a fine collection of pictures, and the other was an inventor who had built himself a forge, in which he had made a boring machine for his own use, a patent shackle to couple trams together, and a wonderful little safety-lamp that could be used to test gas. If the oil-flame was put out by the gas, he had an arrangement for the flame to be relit on the spot by an electric spark inside the lamp. It was a most efficient lamp, very light to handle, and he had made the standards and the gauze himself—in fact, everything except the glass. He was sure that he only lacked someone with influence to take the lamp up to make it a success; anyway, it was the finest lamp I have ever seen. Both of them were enthusiasts at making "Bullets", and I could hear them arranging their lines and arguing about the weaknesses of the winners every week.

On my other side was a man who was always trying to get a tram out of his turn, and as trams were not plentiful there, we did our best to stop him. I remember one day when I had noticed that the usual two lights were in his place and that a lot of talking was going on. I went into that stall unexpectedly, and found that this man was working by himself that day, but had taken his mate's lamp in as well as his own. He was talking to nobody, just to deceive us and get a double share of trams. The haulier had been deceived as well.

Our first pay was due about a week before Christmas, and we were very glad to have a pay that was a decent amount over the minimum wage. In fact, we were delighted to have a pay of any sort. The first day that I saw the measuring done at that colliery the overman and under-manager brought with them another man who was exactly as Father Christmas is depicted. To see a bearded man in the mines is rare, but to see an old man with long white hair and a beard reaching almost to his knees completely astonished me. As it was so near to Christmas, I began to wonder if Father Christmas had forsaken his aerial methods and found a short cut up from Australia. This man, however, carried the measuring-tape, but did not give one inch away.

The boys were interested in this beard. They had a tale that years ago, when this man had been working at the coal, he had driven the top of a lid over a post to tighten it, and while doing so had allowed his beard to be fastened between the post and the lid. As he was the last man in, and had thrown the sledge back, he had to stay there held by his beard for hours, until someone came to look for him and knocked the post out to release him. I did not believe this, but a beard could be awkward in the lower places of the mine.

Perhaps I should explain what is meant by a lid and a post? A post is a piece of timber, either foreign or home-grown, which is set at an angle to meet the slope of the roof. It must be set solidly in a hole at the butt, but is usually measured so that a space of four or five inches is left on top.

Another flattened piece of timber is then driven over the post at right angles. This shorter piece—or lid—tightens the post, and also presents a much greater length of support to the roof.

A week before Christmas a notice was posted outside giving a fortnight's notice to all men employed at that colliery. We were heavy-hearted again at being idle so soon, and on the last day of that fortnight everyone prepared to send their tools out on top of their last tram of coal—filling a small one for that purpose. As I was arranging the lumps on my tram, the overman came into the stall.

"Make it up to a good one," he said.

"I'd best leave room to put the tools on," I argued—"it'll be a lot easier than carrying them out."

"Never mind the tools, then," he said—"leave them here for a bit. There'll be a chance to get them another day if you want them."

I left the tools there, and we lost three days. Then I had a message to say our place was restarting and I was to go to work. I noticed that not more than half the men returned to work, and most of the rest never returned there. It seemed there was a possibility of trouble about the list, and the management used this method of getting clear of the awkward men.

We were working a sort of anthracite, or stone-coal, at that place. It is harder than the steam- or the house-coal. It is of a more shiny nature, and is more difficult to cut. This coal has sharp edges which cut the hands. My hands were festering continually, and it was agony when I started to wash or had to place my hands in very hot water. I caught a hard-driven cricket ball one Saturday afternoon and more than a dozen cuts opened and started to bleed from the jarring.

Often I had tyre-bursts or breakdowns that made me walk all the way home. One day a piece of coal hit me on the thigh. I worked to the end of that shift, then found I had a job to stand on the leg. I hopped down the incline, consoling myself with the thought that it would be easier when I could put my weight on the bicycle seat. I had not ridden two hundred yards when I had a tyre burst—a real gash that I had no hope of mending. I started to walk that seven miles, with every step giving me severe pain, and before I had gone one mile it began to rain; then came the deluge. It maintained that downpour all the way, and before long I was lying on top of the cycle and dragging the injured leg along that way. I did think of lying in the gutter until someone found me; certainly I could not have been any wetter.

Soon after this the first charabanc arrived in the village. It was considered a fine one, and was taken slowly through the streets—with the intention of getting us accustomed to it and for us to see it was safe, I suppose. People ventured to use it after awhile, and their fears that it would go too fast were groundless. After awhile it was thought best to make all arrangements for days in front when any-one went on a trip in it, for there was no saying when one would return. The football team once returned from a distant fixture at tea-time on the Sunday; even then they would have been much later, but the experience of the Rugby forwards in pushing helped their return.

Later they bought a smaller and more efficient charabanc. Many men from Treclewyd had been given work at that colliery, and I believe there were quite thirty of them on each shift. So they booked a charabanc between them to take them to work, as there was no other suitable way. The old-timers were sarcastic about this and complained:—

"They don't know they're born nowadays. They can have motor-cars to take them to work. In our days we had to walk, so we did."

They did ride to work, and had to pay a reduced rate of a shilling a day for the luxury.

It was to save that big item of a shilling a day that I continued to ride my cycle for some weeks, but when I had done a few more walks home, I listened to my mates, and went with them in the charabanc. That day the driver ran into a wall, and we had to walk home again.

I think that was about the beginning of travelling any distance to work from this locality. It has become a recognised thing by to-day. There are good facilities now—double-decker buses and train services—to bring the two thousand and more men who come into this valley to work. They come from a radius of eighteen miles, and pay up to nine shillings a week in fares. If there is no work for them when they arrive, they must return home by some method. They get no pay for that day, but have to pay for the ride. I have known this happen once a week, and sometimes twice, over a period of four months. This travelling means that they are in their dirty clothes for an hour and a half extra at each end of the shift, while in the winter they get nothing to warm them from early morning until evening.

There is another side to this travelling to work. The officials often prefer to engage men from a distance because those men are difficult to organise. They must hurry to their conveyances, and cannot attend a meeting or anything that concerns the work. If they are a shilling or two short in their money it will not pay them to come to see about it. If the committee are local men—as it is safe to assume they must be—then the outsiders do not know them well, and feel that they distrust them because they are living near the colliery and the officials.

These outsiders are compelled to pay towards institutions that they cannot use, and have to allow deductions from their pay for welfare schemes, bands, nursing services, etc., that they live too far from to enjoy. They are brought from the streets of towns like Merthyr and Dowlais or Aberavon, and taken to work in an area that is beautiful in comparison—except the spot where the collieries are built. They have to contribute to the building of sports grounds, parks, and libraries, yet every evening they are taken past them to their drab homes, with no chance of getting enjoyment from these facilities, and every week they must pay the bus fare out of their small wages.

I travelled by charabanc for some months, and, in spite of the distance, I liked working there as much as a man can like working in a colliery. There was little, if any, serious bullying by the officials, and I earned enough above the minimum wage to cover my fare and a bit over.

One of the boys was late one morning, so we sent to his home and found they had slept late, but were hurrying to get ready. Time was getting on, so some of the men wanted to go and leave him behind, but we carried our point that we should wait, because the boy's father was idle and his earnings were all that was going into that house. Finally he came running, with a piece of food in his hand. He was fifteen years old, a quiet and well-behaved boy.

Ten minutes after we started work he came to me asking for a loan of my hatchet for his butty. This butty had not been good at returning things, and sent this boy to ask for a thing that might have been refused to himself. The boy asked so nicely that I could not refuse, but I warned him to bring it back within half an hour, for I should want to use it myself.

"Don't you worry," he assured me, "you'll have it back quicker than that. I'll see that you do."

He never did. They called me there about ten minutes later. By levering with a post they managed to lift the stone high enough to get him from under. I remember how uselessly his head sagged when they tried to lift it. I think that most of his bones were broken.

We had not been away many hours when we brought him back home. As there was no ambulance to be had, we sent for our conveyance. We had to place the stretcher across the top of the seats and sit below it with our hand held up to steady it. Everyone along those miles that we travelled could see what was on that stretcher. There was not one blanket to be had at the colliery, so we covered him with brattice-cloth and our overcoats. We could not drive up to the house, so we carried the burden the last hundred yards, and his father ran down to meet us, calling the boy by name. He collapsed and fainted right at my feet. I have always been glad I loaned my hatchet, or I would have felt that I withheld the thing that might have saved his life.

Soon after this the colliery went slack. We went up day after day and came back without working.

One morning while waiting for our 'bus we noticed a red glare above our driver's garage. We went nearer to see what was on. Our driver had suspected a petrol leak, and going under the 'bus with a naked light, had found it very quickly. He escaped himself, all right, but all that was left of his charabanc would have gone into a wheelbarrow. So we had no comfortable way of going to work any more.

My old working-place at the colliery on the mountain had been working again for some months, but I had not bothered, because I was well suited as it was. But the slack time at the Perry had altered things. I saw the under-manager at the Treclewyd colliery that night and complained that I had not been given my turn to restart; although I knew, which he didn't, that it was my own slackness in inquiring about how many men had started that was to blame. It was settled that I was to start back at the old colliery that night.

There was another chap working not far from me who had himself only recently returned from the same colliery as I had. We had not been at work more than an hour on my first night when a piece of clod fell on him. He was between four posts in a cage, where he had barely room to move his mandril, and there was not a yard between the farthest posts—yet a piece of stone fell between these posts. It did not fall more than eighteen inches, and it weighed only half a hundredweight. It hit him on the back when he was kneeling, and although he was young and strong, he has never walked since.

I knew the coal-cutting machine was working again there, and had kept away because of it. I had since been told that a man from the firm had come to drive it—that was one of the reasons I felt inclined to go back.

So I re-started as a collier. We were three men in each stall, and had to push our trams quite a hundred yards. This was hard work with one man pulling in front and the other two pushing hard with their backs and pressing their feet into the ground. It was not high enough to take a horse in there. We often filled sixteen trams in a shift and pushed them in and out. Not a minute to spare, always working at top speed, then getting very little more than the minimum wage.

After I had started I found that the machine operator had only arranged to stay there for a short while, and was definitely going away within the fortnight. I was the obvious choice, but I felt nervous about tackling the dangers of machine-cutting after being away from it for so long. But when a man has the choice of a job he detests or no job at all, he usually takes the job—I did.

My old mate Tommy had returned to his native Swansea, and his father had got him a job on the docks. He is still there, but I believe that the cottage at Blackpill is as far from becoming his as ever it was. Tommy had been hurt at work, and his father came to see him, so after he recovered he didn't want to go back underground to work.

Tommy's father decided to come and see his son on a Saturday after work, and started to walk there about tea-time. He was a lusty man, a good walker, who was contemptuous of anyone who was not so good a walker as himself. He had no use for trains, and considered twenty miles a decent stroll. So it was only a little amble up to our village. He was soon half-way there, and was getting along so rapidly that he thought he had plenty of time to sample the ale at one or two of the roadside inns. The evening was warm, and he did not travel so well on a full stomach, so when he was not more than a mile away from his son's home, he sat on the grass bank at the roadside and went to sleep.

A young man with a motor combination came along, and it broke down right by the sleeper. After awhile Tommy's father awoke and took a vague and abusive interest in the repairs. He was sound asleep again before the machine was properly right, but the motorist—who had found out that the other came from Swansea—did not like to leave him lying in the dark alongside the road, and struggled to lift the sleepy one into the sidecar. Tommy's father had some idea that he was being given a lift somewhere, but he did not get properly awake until they got back into Swansea. What he then said to the man who thought he had done him a good turn attracted a policeman and repelled the magistrate.

July 4, 1924, was a great day for me as well as for the Americans. That afternoon I went upstairs to see my son—Peter—having his first look at this world. I can still remember thinking how small and helpless he looked then, but he has been busy remedying these defects ever since. One day old, I thought; it will be a terrible long time before he's even a year old, and it is ridiculous to think of the time when he will be able to walk and talk. Time, however, seems to flash along, and those desirable events came very quickly.

On the machine in place of Tommy we had another Englishman named Billy Ward. Many years before that he had been a tram-driver, and his wife the conductress, but when I knew him first he had six children, and was having a rough time, so that he welcomed the extra time we worked on the machine. He was a very intelligent man who wanted to know all about everything. He had an under-manager's certificate, and was a class cornet player. He was very fond of his children, and was often making toys out of wire or rags for them because he could not afford to buy them. After a spell of hard work he would go and fetch his pay, and then go into a public-house to have one drink. Instead of only one he would keep on drinking. He was disgusted with himself afterwards, and used to tell me how ashamed he was of having spent the money on drink instead of getting things for his children. I suggested that he should give up drink altogether if he could not control himself. He told me that his mother had to go out to work when he was a tiny baby and take him with her. The work she got—so that they could live—was cleaning up at a public-house, and when he was not old enough to walk he used to sup at the slops to satisfy his hunger. He blamed his childhood for the craving that the smell of beer gave to him. He was much more nervous than the rest of us, and one reason was that he had been shell-shocked, and badly injured in the mine some time before.

Hutch did not make him any better, for he delighted in worrying Billy. If we were under a bad piece of top, Hutch would pretend to argue that Billy ought to go under there because he would have more compensation, as he had so many more children than Hutch We were again working long hours, and after perhaps eight hours without food, Hutch would insist on giving Billy luscious details of great feeds he had eaten in the past—complete with cream slices, chocolate éclairs, and other delicacies.

Ward was a Spiritualist, and said he heard strange sounds or saw unusual things. Sometimes when we called for him to go to work he would definitely refuse to come because he had heard "them bells".

"No, lads," he would state—and we could never change his decision—"I heard them bells a'ringing again last night, and when I hear them there's something going to happen. I'm not coming, so there."

So we had to go without him and defy the ringing of the bells. Several times I have found him waiting for our call of the evening with an old music-stand in front of him and his family sitting around while he made that old cornet sing in spite of the wire and rag that were holding it from falling apart.

He used to alarm me sometimes with his warnings. I remember when we were cutting under a particularly bad piece of roof and had stayed for a few minutes to snatch a bit of food. When we got up again he walked up to me and said:—

"He's gone now, but you'd best be careful."

"Gone?" I was puzzled. "Who d'you mean?"

"Ah, lad," he said, "you only thought as there was four of us there while we was having grub. But I could see another one. He was stiff-built like you, and looked like you in the face, and he was looking at you all the time. When you got up he faded away. I don't think it means anything bad for you by his look, but you'd best be careful."

His insistence that a ghost was following me about was not consoling when we had another six hours of dangerous work, but I never made fun of Billy.

I try to respect every man's opinion; besides, I was fond in a way of him.

One morning in 1926 I watched Peter achieve his first solitary stroll up the garden path. He was very brave over it, but when about half-way he swayed and caught at the stem of a flower for support. Then he seemed to realise the beauty of the flower, and moved his head back and forth in his eagerness to view it from all angles. He did not damage it, but left it nodding while he resumed his swaying journey. He looked back once at the flower and smiled at it, as if he thought, " You're the same as me, aren't you?—just doing your little best to cheer this grey, dusty place up a bit."

We were paying a small amount back every week on the money we had been loaned from the parish; we were behind a great deal with the rent; and we had shopkeepers to clear and worn clothing that needed replacing—then, before we had properly started on these repayments, another stoppage was threatened.

Our hopes became fainter as the days passed with no settlement, and we were warned to put the working-place as secure as possible in case of a long stoppage.

We had a worrying hour at the end of the last shift before that stoppage. The running-cable had been cut short and would not reach to the next switch, so we managed to cut on by attaching two weaker cables and going very slowly. This was dangerous, for there was a big risk of being electrocuted, and in that last twenty yards we went under four " bells ".

A bell—in mining language—is a stone which is bell-shaped, with the widest part down. It is a smooth, circular slip, which gets smaller as it goes higher. Because of this shape it is liable to drop out at any moment, and so is dreaded by the workmen. We could not put posts under them, or it would have blocked the machine-way, so we had to dodge slowly along and trust to the luck that seemed to follow us—mostly. It stayed as our helper too, and we needed it, with four death-traps above, over-strained cables that might burst at any minute under us, and the prospect of idleness before us when we did get through.

"There! She's safe for a day or two, I suppose," I said when we switched off after reaching the top of the heading.

"P'raps so," said Hutch, "but I think that we'll be here again to-morrow."

We hoped he was right as we coiled the cables on one side and put the tools safe; but he was not. It was many months before I handled that machine again.

CHAPTER TEN

WE were in poorer shape for a stoppage in 1926 than we had been in 1921. None of us had any reserve funds, and few had paid much back off the debts. We had carried all the old timber that was to be found on the mountains, and the coal-tips had been picked as clean as possible. The old pits we had used to get coal from had been blown in to stop anyone going down. This was a wise precaution, for the men would have kept on risking their lives for that coal as long as they were able to get down.

We were proud of the solid front of the Unions at the beginning, and were confident that we would have their support until we gained a fair settlement. We changed that opinion after a few days. Ever since I have felt that either our leaders betrayed us or they were too chicken-hearted to do in a difficult time the job they had done in easier days and had been well paid for doing. I think that, even until to-day, the great majority of the men blame our leaders for what happened, and I have never heard one of my workmates speak a good word for J. H. Thomas since that day. A. J. Cook was the only leader who seemed to have kept the respect of the men after 1926.

The working miners showed that they were not prepared to give way by the determination with which they maintained the strike. Even at the end the majority of votes were against surrender.

All through that summer the stoppage was continued. We walked up the mountains because the grass was soft and because we could not endure the sight of those prosperous-looking cars flashing along. Weeks of waiting and hoping merged into more weeks when we just waited. This dispute had gone on so long that we began to get indifferent as to when we would start, for it seemed that we must have another drop in wages and a return to longer hours. We had forgotten what it was to have spending-money, and men tried all sorts of tricks to satisfy their craving for a smoke. They tried cutting herbs and smoking them. A cigarette stump in the gutter would cause a rush to get it first. Men went long distances, dodged the police, and hunted on the distant tips for hours so that they could get a sack of waste coal which they sold secretly for eighteen pence. Parcels of clothes came occasionally, but it was usually the loudest shouter, not the most needy, who gained by these. The ones who are in real need do not often go out on the streets to broadcast their poverty; that is why the superficial observer rarely learns what he sets out to do.

I did not waste all my time during that summer. They started to give lectures under the St. John's Ambulance Association at Treclewyd during that time, and I attended, as did more than sixty others, for there was little else to interest us. A young doctor fresh from college gave us a lecture, and aired all the difficult words and terms he had ever heard. After that we never had twenty in a class, but it took more than long words to frighten those who remained. Week after week we gave the only coppers we had to rent a room and buy bandages. We studied the text-book until we knew it by heart.

We went under a tree on the mountain-side and bandaged each other up. We were grateful to anyone who showed us where their bones had been fractured and set; and we walked miles in each direction to attend lectures. After we had passed our first certificate we changed our usual walk one night so that we could be near the spot where an accident had happened the night before, because we reasoned that these things always occur in twos or threes. When we returned home we found that a smash had happened near our usual haunt, and we were blamed for not being there.

We borrowed bicycles and cycled more than ten miles to an important Ambulance Competition. Although we had no uniform and had to borrow equipment, we won the first prize with a margin, and out of that five pounds we bought our first stretcher, splints, and blankets. In the autumn I had an idea for funds which resulted in faking a car smash on the side of the road—with stretchers and bandaged patients—and we collected off the cars that were stopped to see what had happened. We had three enjoyable days, and the funds were richer by over sixteen pounds.

I made a new friend during that summer: a young English doctor who had come there as an assistant and had set up on his own. He had arrived there some time before I knew him, and had been disgusted with our village, our surgery, and our music—especially the surgery and the music, for he was a musician—a real one. He doubted whether it was worth bringing his violins there for the short time that he intended to stay; but felt so lonely without them that he decided to risk contaminating the lovely instruments.

One night he went to a concert, fully prepared to

suffer. The fates were sniggering at him that night, for our best pianoforte soloist was playing, and her artistry made him forget his dislike of what was outside. Dusty streets; grey and overcrowded houses; injured men with black dust covering their broken bodies; ailing children needing better food to build up their strength, and worn-out mothers who were tired of trying to make a little money bridge the gap between one pay day and the next—all these things faded from his mind as he watched those white fingers dancing along the keys of the concert grand piano.

She had a wonderful touch, had that young lady, and a face, figure, and disposition that matched her music-making. He had the ability to go far as a player, but I doubt if he would ever have been better rewarded. He talked no more of an early departure, but instantly decided that the village that could produce such a musician was honouring him by letting its dust dirty his spats.

They play more duets than solos now, and I know that discords in their married life are as rare as they were in their music. I got to know this doctor during 1926, and he taught me a deal about physiology and anatomy. I could not help learning music from him, because anyone who was in contact with him was forced to do so. I used to sit for hours in the evenings listening to his playing, not daring to move my position in case he would stop. He loaned me an old violin, and I started having lessons with him. I had plenty of time to practise hard, and I was to pay for the lessons when I was working again. I learned—slowly, I admit, but very surely—and when the time came, I paid.

I was again taking my place in the relief queue.

Six deep, it stretched all round the large chapel and along the road. There was no privacy inside or out—we were on show for everyone to see, although it did not matter much in this village, where ninety per cent. of the men went there. Everyone could hear what was said inside. Some of the men did not mind, but the more sensitive ones suffered a great deal. The ones who stammered seemed to feel it badly, for their nerves were upset by the conditions, and there was an impatient officer and an eager crowd waiting for each word. I have often thought this a method that needed alteration at the Labour Exchange as well.

One week, when we were waiting our turns, I noticed our chairman calling the committee out of the queue. It was obvious that they were worried over some development, and we were warned to attend a general meeting that night. Then we found what was troubling them. The company that employed us had sent for the committee and told them that they found the colliery was getting into a bad state of repair because of the long stoppage. They wanted to reopen it, and were willing to pay the wage that had been paid before the demand for a reduction, with an additional bonus for every man that went to work.

There was no question of what the men would do: they turned down the suggestion at once. They could not break away from the rest of the Federation and disobey its leaders; and that offer of extra money was one that every colliery company was prepared to make if they could have their men back at work in the time when coal was so scarce and dear. No one raised the suggestion in the meeting that it would be wise to return to work, but we noticed two men, whom we had reason to suspect as tale-carriers, leave before it was properly finished, and we were warned that they had gone straight to the manager's house.

Next morning these two men started work. We were surprised, but not greatly concerned. Two men, even if helped in some degree by the officials, cannot work a colliery. Some of the committee tried to dissuade these men, but failed, and were abused by these two men, who appeared to be anxious to force our committee-men to say something that could be construed into a threat.

A week later two large lorries arrived loaded with men from a town about twenty miles away. These strangers walked up to the colliery, well guarded by policemen. There were about sixty of these blacklegs, but not ten of them were real miners. The novices had to be put to work in safe places, while the more experienced men looked after their safety. They did just as much work as they pleased—which was rarely a great amount; yet the colliery-owners, who had been insisting that our wages must be reduced to less than ten shillings a day, were willing to give these blacklegs fifteen shillings a day and to pay all travelling costs.

They brought no tools with them, because those bought by the real miners were still in the colliery. The locks on these were forced and the tools used. As the new-comers had no skill in handling them, they soon broke the handles and blades, for this is often more of a sign of clumsiness than strength. When a handle was broken, the mandril or sledge was thrown into the gob and buried. There was a good supply available, for the tools of over four hundred men were inside waiting to be stolen or smashed. Several times I saw a hatchet, the handle of which had been broken, carried out by the blacklegs as they went home, and the picketing miners would see their own tools carried away past their noses, yet were prevented by the police from interfering. I saw one blackleg taking the hatchet with him and shouting to the others as they passed the pickets, " This'll do me nice for chopping up the sticks at home." That tool would have been worth at least twelve shillings, and the man to whom it belonged was standing watching, but if he had shouted, " That's mine ", he would probably have been arrested. We had been warned not to make a sound or give the least excuse for the police to interfere.

On rare occasions, when the blacklegs felt the instincts of a normal man, the hatchet or mandril head would be sent to the screens on top of a tram of coal, and one of the committee would have to ask permission to fetch it and restore it to its owner, so that he would have it repaired when he could afford a stick.

A miner must have tools with which to work, and a set would cost, at the cheapest, three or four pounds, so it was galling to the local men to see the blacklegs coming there to take their living from them and grinning in their faces as well when they carried the tools.

It is one thing to fight on with determination while all the men are idle, but it is another when the work has started and you are idle. There is the prospect that your place will be filled and that you may be idle for months, and perhaps for ever, after the dispute is over. You wonder every day if some of the others are going to go back to work, and if you will be left out to be penalised in the end. There is also the need to answer the relieving officer very carefully when he snarls at you that your colliery is going and some men are getting good money there; so how is it you are idle?

It is also a remarkable fact that there are men who will not work when the collieries are going full swing, yet suddenly discover a conscience during a strike, and proclaim that no one has a right to stop them working. I knew one man here who had most of the attributes of a class in society that was not his own. I think his ancestry should have been interesting. Year after year he refused to do any real work; he simply wore good clothes well, although how he obtained them from the parish relief and other doubtful ways of getting a living he used I could never make out. He had a lordly manner; instructed everyone how to work; flourished a walking-stick; and did not smoke anything but cigars—I often got the smell of one from him. He would have made a most efficient lord. He was true to type at strike-time, for he insisted that no one had a right to prevent him working, and away he went blacklegging.

Every day we walked up the first slope of the mountain to meet the blacklegs coming from work. We stood silent, not even looking as if we wished these interlopers had to be carried out instead of walking. We waited near the top of a gully that was quite eighty feet deep, and there was a churchyard handy as well, but all these natural advantages were spoiled by the continual presence of imported police. We made sure, however, that no one ventured to work during the night time.

It seems that the present law has made picketing a farce, for of what use is it to appeal to the morals or principles of the type of men whom I saw there as blacklegs?

Week after week went by, and a small amount of coal came down every day. The men and their wives and children stood on their doorsteps and watched the "journeys" go by, while they were hungry and penniless and their own tools were being used, but there was no talk of going back to work until all went. They were amazing in their determination, the women as well as the men. The company spread stories of big money being earned by those who worked, but the regular workmen remained penniless, and loyal. All through those three months not a dozen local men broke away. The company got some coal out, but paid dearly for it.

They tried to arrange so that a skilled man should work with one or two of the novices, as far as they could, and the officials themselves had to post the stalls and sometimes use the mandrils. I was told that they were having a lot of trouble with the coal-cutting machine, and that the under-manager was handling it himself—with such results that he would have been very drastic to us had we done no better. In spite of all their care, several of the blacklegs were injured; and there were no demonstrations of sympathy.

The unemployed were better off then than the men who were concerned in the dispute, for they did have a little money coming to them each week. The only inducement that could persuade them to blackleg was the chance that they might gain regular work after the dispute was settled, and this promise was used to the limit. There were several hundreds of unemployed here then, yet only one was persuaded. He did start, and the Labour Exchange seized the chance to threaten to stop the pay of the others unless they followed his example.

The men refused to be stampeded, and they won their case—the pay was maintained.

There was one man here, with a reputation of being simple, who went to the colliery manager early in the stoppage and asked to be given work. The colliery was completely closed, and the manager could not consider opening it for one man, and such a man. That refusal was just what this appellant wanted, for his dole had been stopped; because he was unemployable, I suppose. He then claimed that it should be restored to him, as he had tried for work and failed to get it. He made out quite a strong case, did this fellow who was supposed to be a simpleton—good enough to defeat the local officials—and the demand had to go to a higher tribunal. I never heard whether he won or lost his case.

Some weeks before the end of the stoppage I heard that Billy Ward had gone to work. He was living in an isolated cottage on the mountain side, and the under-manager had been there persuading Billy that if he did not start he would have no chance of work when the stoppage was over. Billy did not have the support of other men living near, as we did, and I suppose the thought of his wife and children helped this threat of the official, who was very anxious to get some of the machine-men back to take the job from his own shoulders. I felt this action of Billy's very much, because I suspected that he would be put in charge of the cutter, and so I should not have such a good chance of getting back.

Then came the time when we were told that work was to be resumed throughout the coalfields. We were to work longer hours for little more than half the wages, while all the consolation we could find was that we had made a good fight of it.

We had our problem as to what was going to happen to us, for we had been warned for months that those who did not begin work would not be taken on after a settlement. If our places were filled, where could we hope for work, for most of the other collieries would be many months before they were in good enough repair to engage all their old workers?

We had a general meeting, then waited on the road while the committee went to the office to see the manager. They wanted to know what was the position?—were the old workmen to be re-engaged? He did not allow them to step inside the office out of the rain. He opened the door so that they could see the police waiting there, then he cursed them for all he was worth, brave in the knowledge that they dare not answer back.

He would not allow us to sign on as willing to start when work came, neither would he give any assurances that the old hands would be taken on when work was available. They went back to him again with an appeal to give particulars of our case to the Labour Exchange so that we could put in a claim for the dole, but all he did was tell them to go to hell from there. The police appeared to enjoy this baiting of anxious men.

We dawdled our way home, about four hundred of us, and the future looked very black for us all. Sometime, about eight o'clock that night, I went to answer a knock at the door, and found the under-manager there with instructions to go to work next morning. He told me that he was mighty glad the dispute was over, and that he was hoping he would soon get the old hands back to replace those useless articles that were there at that time. Half a dozen of the old repairers were to begin work, and the other two machine-men.

I was elated by this news which eased my problems, but I went to ask our chairman if it was right for me to start, because I had no intention of leaving my mates in the lurch, no matter what it cost me. He was pleased that we had been given the chance, because he reasoned that the more of the old hands that were reinstated the better chance he and the committee had of getting back. It seems to me to be a most unfair thing to penalise the men's officials after a strike. They only carry out the mandate of the men they represent, and this committee had not been a warlike one by any means.

I was at work again next morning. Nine of us re-started, with more than a hundred blacklegs—and what a selection these latter were!—the dregs of a large town coming to take advantage of the distress of their fellow-men and help to starve their women and children; accepting a bribe to do work that was too hard and dirty for their liking when things were normal, and quite content to return to their leaning against street corners after they had done their share in lowering the standard of living.

The nine of us walked up together and kept to ourselves. Billy Ward came across to us, but we did not speak, and he fidgeted alongside us like a dog that wanted to be forgiven for a mistake he had made. We were on fire when we heard the blacklegs boasting of what they would like to do to the men who were picketing and the gentle treatment that the women would receive if they had the least chance. Billy did not go near them

—he stayed by us, hoping for the time when we would be the same to him as we had been in the old days.

I was again put in charge of the cutting, and Billy was to help behind, as he had not been a success as a driver. We travelled along the coal-face that day, and the days which followed, without speaking a friendly word to the workers in each stall we cut. We made no effort to put things right for them, only did exactly what we had to do. This was in contrast to the old days, when friendly banter was exchanged as we cut past each group of men, and when we worked together with them to post the face safely or clear coal out of the way. We used every opportunity to rescue the tools belonging to the men that were idle and hide them ready for another day.

We walked home in a group that kept a distance behind the others, and when the pickets had watched them pass with lowered heads, they closed in around us and we walked together—nine black men amongst some hundreds of white-faced pickets.

We were prepared for a shock on pay day but, even then, the figures on the docket completely staggered us. That is one thing that has always puzzled me—how do the newspapers arrive at the figures of our wages? When we have a rise it is never such a large one as the Press figures lead us to expect, and the lowering of wages is always greater after a reduction than they estimate. It may be done with the intention of easing the blow to us, but I am sure that people who are not in the industry are misled in this way.

We had one consolation on that first pay day. It was pleasing to our feelings to listen to the clamour of the blacklegs, for they were infuriated that they had now to work for less than ten shillings a day, instead of the fifteen shillings they had been given while the dispute was on; also the lorry rides were to be free no longer—so they were informed. Several of them loudly declared their intention of putting their working boots in the manager's face and of never coming near that embroidered hole again. We did not try to dissuade them from keeping either of these resolutions—more than twenty did carry out the last one. After that week everyone was expected to do the requisite amount of work, or he was not paid, and the number of blacklegs dwindled rapidly, until none were there, of the travellers, in two months' time. We welcomed each one of the old hands back like a confidence man greeting a prospective victim.

We attempted to re-start the payments to the Federation, but the manager had his way in that matter, and he would not allow the chairman or the check-weigher to come on the colliery property. The check-weigher there was quite a mild-mannered man and had been severely criticised at a general meeting because he had spoken to the manager in the street during the dispute—yet he had to suffer. He is the fourth check-weigher—elected to see that the men get fair play in the coal-weighing—I have known whom the colliery companies have dismissed. We built a small cabin just outside the company's premises and paid our Union dues there.

As the Federation was not recognised and we elected no men to represent us, because of that action we were compelled to take almost what payment the owners decided to give us, and we had no hope of appeal. As they did not interfere with my wages, I kept very quiet and continued with the work.

We went on in this way until February 1928. We worked on late on Saturday morning, and when we saw the under-manager outside he warned us to be on time the next morning, Sunday. That night a notice was put out in a shop window that there would be no more work until further notice. That was all the warning four hundred men had that they were to be thrown on the scrap-heap.

It was rumoured that we would soon re-start, but the weeks passed and we were still idle. We understood that the partners in the firm had quarrelled and that the chief partner was determined to retire and leave the works closed up. This dispute could not be settled, and when some men went to the royalty-owner to ask him to reconsider his view on the payment of royalty in the hope that the owners would reopen the colliery if he did, they were made to feel sorry they had appealed.

So the coal remains under the mountain untouched, while many of the men have not worked since that time, and get what little satisfaction they can from the thought that one man no longer draws a lot of royalty money whether he wakes or sleeps.

The engines and other machinery were blown to pieces because they could not be taken out of the buildings any other way, and the buildings had to be left standing; they were sold at scrap-iron price, but the coal-cutter was taken to another colliery. I often wonder why the Miners' Federation or the Co-operative Movement doesn't buy some of these mining concerns that are sold at scrap-iron prices. There are plenty of young, and qualified, men in the Federation who would welcome the chance to show that these derelict mines can be made to pay. It would show that the present method is wasteful and wrong.

About a fortnight before the stoppage I had been hit very hard on the left side of the neck by a post that fled out when the rope tightened against it. I had been in some pain for days, but had kept on working under the impression that it was a severe bruise and would soon ease. It had not been reported because knocks of this kind are so frequent that I felt ashamed of making a fuss concerning it. Yet the fact that it had not been entered in a report prevented me from having any claim for compensation when it began to get more serious. The side of my neck kept swelling up, and I had pain every time I ate anything or swallowed.

I was again very fortunate in not being idle for long, due once again to my knowledge of machinery and my ability to combine the work of a miner with that of a mechanic. The other miners had a hopeless quest, for nearly all the collieries around there had lists of their own men that they had to start before they could engage outsiders. These men from my own colliery had no claim on any management for seniority or re-engagement. They were completely against the wall, and there seemed little hope that they would ever get work again. I remember I took one of them with me when I resumed the hunt for work. One manager, who knew us slightly, said he had a hundred of his own colliers waiting to start, but as he had intentions of installing machines there, he would start me, alone, so that I could prepare the place and handle the machine when it came.

I had my first experience of seeing the sky jump away and the solid earth rush upwards as I dropped downwards when I went down that pit. Once at the bottom it was quite the same as the drifts and levels. My work was to straighten out and reopen some old workings that had once been an iron-ore mine. In the airway I found an old grate with some coal and sticks laid alongside, as they had been left over a hundred years before by the man who had last been through there before me. The method of ventilating this type of workings in those olden days was to light a fire, and as the fire would draw the air towards it, this would induce a current of air to move and ventilate the workings. I expect it did not give very good ventilation that way, but I found the present ventilation was little better, for I was always afraid of being overcome by foul atmosphere amongst those old workings where I spent most of the time by myself. It was a part of this colliery that was some distance from the real workings, and we were trying to open up to a seam that had not been worked. One of my duties was to hurry away to where a small auxiliary fan had been placed if I noticed a light approaching that I thought might belong to an inspector of mines. I don't think that my turning of the handle of this tiny fan made much of a movement amongst the atmosphere of these workings, but the object was to show that a fan had been put there and that some attempt was being made to improve things.

I had to travel some distance from home, and most of the time I was at work I was suffering from my neck. I drove a short pit down about twenty feet and found the seam. It was only eighteen inches thick. I don't ever remember such a period of agony and misery as I endured during those weeks while I was at work there. I had to lie flat and force my body along the coal-face. Where the roof was weak I had the pleasant realisation that if a stone dropped two inches I must be smothered before help could come. All the coal had to be blown out with powder, and the seam was so thin that I had to bore five holes before I had enough coal to fill one tram. It was quite a contract to bore even one hole. I had to lie flat and force the drill inwards with the pressure of my stomach, while my hands knocked against the top and then the bottom when I turned the handle. I had to move about by drawing myself on my elbows as would a man swimming. Fourteen inches was the length of a post to support the roof, and while I was squirming about I was hitting my head or shoulders against stones and posts that I could not see because the ventilation was not good enough to clear the powder smoke.

I should gladly prefer three months in gaol, if I had the choice, to another three months under those conditions. There were days when I hoped that an accident would happen to me so that I would be crippled enough not to have to work there again. At the end of three months they decided that it would not be suitable to use machines there for awhile, and I was unemployed once again, and could have been sorrier.

So I was back again amongst the ones who walked the valley bothering the managers. I soon learned that there was no chance of getting work as a collier, so after I had been refused a start at any colliery, I asked if they intended using machines there, and showed my references and the letter of congratulation I had received from the coal-cutter company after a long period of record cutting. One manager told me he was having machines shortly and he took my name and address. He told me that I should start as soon as the machines came and that I need not worry about seeing him, he would let me know. I dared not rest on that assurance, however, and I walked all the way up there Monday morning after Monday morning, partly so that he should remember me, partly because I wanted to get away from the depression of the village, and partly because I wanted to keep my eye on that colliery and wondered whether I should ever work there, for it was one of the most modern in the coalfield, and was like a mechanical village—a very busy one—outside.

This walking eased my mind about the home problem and the thoughts of my neck, for I was waiting my bed for an operation. On the morning when I had the paper to go into hospital I went up to see the manager, was told that the machine had not arrived, then went back and away by train to hospital.

I had by that time gained my third-year ambulance medallion and a nursing certificate, so I was interested in what went on in the hospital. In there I was allowed to walk about part of the day, and tried to help the staff. I remember I always had the job of explaining to any new arrival that when a nurse went up to him with a book and asked "Yes" or "No", she wanted to know whether or not his bowels had been opened that day. This message had to be translated to the men in the terms they understood, and these varied. They had a rule that no urine-bottles should be in use after eight in the night, but as the patients needed them badly, I used to smuggle them in and out. They could scarcely dismiss me if they caught me breaking the rules. I used to count the hours through the long night and watch the night-nurses writing up their lecture reports—and sometimes listen and criticise their efforts when they had finished them. An old man opposite me in that ward coughed continually night and day.

Soon after four in the morning, when they started to wash the patients, the nurses would disturb me—and complain that I had kicked the bedclothes all over the place—to make the tea for the ward. The first morning I made the best cup of tea ever drunk in that ward—and used two days' allowance of tea.

I went quietly past the screens that were around the bed at the top end of that ward one morning to see if the occupant there needed tea. Another case was in there—brought in while I had slept. He was an old mate of mine, who had been brought there straight after an accident in work, where he had lost both legs. Two of his brothers were with him, sitting on each side—still in their pit clothes. They showed their recognition of me by just nodding, then sat on silently, as their brother lived his last hour.

I got at the case-sheets above my bed, and was quite interested in the doctor's reports—although this had to be done on the sly.

On the day that they prepared me for my operation I watched the old trolley, wheeled by two porters, rattle and bang past me as it fetched four others from that ward. It made almost as much noise as would a coal-tram. When I had watched the other four brought back unconscious and rolled on to their beds, the doctors came to have another look at my throat. After a lot of feeling and listening and whispering I saw the doctors go away, then a nurse came to tell me that there was to be no operation on me that day and I could have food.

Three days later almost the same thing happened, and I was then told that they were not going to operate, and that I would be going home in a few days. After I had got back home our local doctor hinted that they thought it was too dangerous to operate because of the nearness to the carotid artery. Anyway, I never had an operation, and have not since; but I was weak for a while after that hospital visit.

I went to see the manager the next day, but the machines had not arrived, although they would be coming before many more weeks had passed.

I had been attending at the Labour Exchange when I was not in hospital, and had given my experience as a cutter man in addition to my capabilities as a collier. One Saturday morning one of the clerks called me and told me to go upstairs to see the manager. He told me that he had heard of a vacancy for a cutter-driver at a colliery in the Western valleys, and that he wanted two references to send there. I brought the one from my last working place. It was such a flattering one that I felt ashamed to hand it over, and it gave every credit to me for keeping a good output during the years the machine had been working. It spoke of my ability as a workman during the time that colliery had worked, over the last twelve years—but it was not enough; they—the exchange officials—wanted two references. I argued that I did not want to bother anyone else for a reference when that one covered twelve years and was the only one that I could get that concerned machine-driving. They must have two. While I waited at the counter I saw another man apply for a job up in England. He handed in two references, one for six months and the other for four months;

and the officials were satisfied. Yet one for twelve years was not enough, and so I went out to an official who had never seen a machine and got another—which solved the difficulty.

This job was at a colliery where the same type of seam was being worked and a similar make of machine being used as the ones to which I had become accustomed, while the Exchange officials warned me that the need for a man had been sent to them as urgent, so that I held myself ready to go at once. Three weeks and some odd days after this the Exchange told me they wanted fresh copies of the references, as the others had been mislaid.

Another three weeks went by, and I visited my manager of hope every week. On the fourth Monday morning I was told that I was successful in getting this job away as a machine-man and was expected to travel that night. I told them I was grateful for their help, but that I had been told to start nearer home that night. This was correct—the machine had arrived, and the manager kept his word.

"Anyway," the exchange clerk told me, "you'd better give me full particulars, and we'll enter that you obtained work from this office."

I shan't forget that first shift at this new place as long as I live. It was a very fine colliery, with immense engines and every modern contrivance to hurry the output of coal. They did not handle one tram at a time here, they rushed them out in strings of forty, which were instantly replaced by forty empties. Fine compressor engines forced the compressed air inside—by means of pipes that a man could crawl through—to drive the thirty-five engines that worked inside the colliery. Twelve hundred men worked there, and the rush of coal-getting went on all the hours of the day. The main roadway and drift, which went straight in for over a mile, were almost as high and safe as a railway tunnel, for they were steel-arched and brick-walled.

The manager and under-manager were keen men, but reasonable, and I found them anxious to do right by the men—as far as it was in their power—and quite as ready to believe a man as an official, when it came to a dispute. But here, as in the other collieries, most of the upsets seemed to be caused by the firemen or under-officials who had had a slight promotion and appeared to be in danger of bursting because of it.

I have known men who change their manner of dressing, of thinking, of speaking as soon as the company selects them for the position of fireman, which carries some responsibility as well as extra pay, a guaranteed week, and paid holidays with it. I have—of recent years—had a deal of experience concerning the sergeants and sergeant-majors in the Regular Army, and I found them more pleasant men than two-thirds of the colliery officials. It must be remembered that experience is not the chief consideration when a fireman is appointed, for a man with ten years' experience is often employed telling a man with three times his experience—and often intelligence—what he ought to do; and on those instructions depend the lives of that man and others who follow behind him.

It must be the thought of their authority that drives these poorly balanced men out of their old selves and makes them domineering autocrats. I have seen them stand on the street corners and shout across at men, " Remember you're not late to-night, or you'll be going back," or, "I shall want you to put a move on to-night, remember." The whole idea seemed to be that they must humiliate these men and show themselves as being in authority to all who are about. I have met many decent officials, but somehow they don't seem to keep their jobs long, for it seems far easier to serve God and Mammon than to work for the coal-owners and be just to the men. The test for a colliery fireman cannot elevate them, surely, for any intelligent schoolboy could easily pass the requirements in English and mathematics; and the gas test is simple—in fact, I think that every miner ought to pass these tests, for his own safety.

Distaste for the things he would have to do prevents many of the finest men taking an official's job. The chief things that are required of a fireman is that he close his eyes on certain occasions and that he always get as much coal out as is humanly possible. Certainly, he must sign a report book every day, but as the same words are used, he ought to be almost able to do that in his sleep. On the line opposite "Gas?" he need only write "None", then opposite "Timber?" he must put "Plentiful". I did hear of a flustered fireman who wrote plentiful on the gas line and none on the timber line. He slipped on to the wrong line, but probably wrote what was the truth for once.

I was not in a fit state to stand the roar and dust of starting that machine, for I was still weak and in pain. The coal-face was very low in that part, and I had been used to electricity, so I found that compressed air was more dusty and noisy. Before end of that shift I could not stand up, and was crawling around on burning knees.

Thirst, too! I was so thirsty that I drank all my

water and all I could beg from the others, then sucked a small piece of coal in the hope of easing my craving. We could not finish in one shift, so that day of agony went on into the second shift, and was not ended until two o'clock the next afternoon—by which time I could hardly stagger home. I didn't want food when I got there, only drink—and rest. By a lucky chance they failed to clear the coal, so that I had an extra day off to recover, and things were not quite so bad after that.

I was working too late to sign on during the first week, although I took my cards every day. When we did finish to time I went to sign—and found that the rats had eaten most of the two cards. They had been swarming close to where we were at work, and they seemed to like the taste of stamps, for they had chewed that part which had held the stamps and left me the other side. Then started a long and wearying series of letters to the Friendly Society's London offices on the subject as to whether rats existed in mines and if, in case they did exist, they were fond of the current insurance stamps. When I ultimately convinced them that rats were there, they wanted to know why I hadn't had the card franked at the Exchange. I told them that there had been no need to get that part franked, as stamps were on it. Then they wanted to know, for the twentieth time, where the stamps were. That was difficult to answer, because the rats had left no trace of the stamps or of that half of the card. After they had amused themselves and annoyed me for nearly three months, they agreed to allow me the benefit of the doubt, but warned me, as a parting shot, to make sure the cards were franked in future. So I think I ought to get the unemployed franking put on top of my stamped card in future, on the understanding that rats do not like the ink they use in the Labour Exchange.

It was the usual tale of machine-mining here again. Rush the machine through, then rush the conveyor alongside. Rush the men to clear the coal and the machine to cut more. Drop the price to the lowest possible, so that the men must work to the limit amongst the rattle and dust of the machines. Rush—rush—rush; and then came the usual end: the roof came in and all were on stop.

It seemed that there would be no more machine-cutting for months in that area, so we were sent to work in another seam. This was the nine-feet, and we were there as repairers. There was plenty of risk on this job, for we had to put the place safe after falls, and were often underneath places where it had fallen to a depth of twenty feet and more; but we had plenty of height to work in, had a good supply of air to breathe, and knew that we were fairly sure to finish on time. It was a relief after the other work, and I determined not to go back if I could avoid doing so.

Sometimes the coal became as thick as fifteen feet in this district, and was like a hill in front of us, but the price was low, because cutting was easy. There was a lot of gas everywhere there, and if I stepped three yards away from one of the currents of air, I could taste and feel the heat of the coal-gas. This district was being enlarged, and to conform with the Mines Act they made a cabin to hold the ambulance equipment, such as a bed, splints, and stretchers. This colliery was well equipped to deal with injuries. It had a good motor ambulance, a well-built and equipped room outside for treatment of injured men, and cabins inside in each district for an ambulance man to attend to his cases.

This was the only colliery I had worked at which took this amount of trouble over its injured.

They were careful that their ambulance men were efficient and in regular training. When they needed one for that district, I was the only one qualified, and so I had the job.

CHAPTER ELEVEN

I HAD now to carry a large tin box strapped on my back, with bandages, lint, iodine, and a tourniquet inside. I had a heavy electric lamp as well, which weighed about fourteen pounds, so that with my clothes and food-tin, in addition to a big jack to hold water, I had a load which fetched the sweat out of me in the warm workings. I always carried a coat in the winter time to lessen the chill when I came out to the cold air. I found it dangerous to get on the damp ground to garden, for that would surely lead to a severe cold, if nothing worse.

My work now was first to see that all the ambulance materials were in my box or available in the cabin. To see that the blankets were there and the 'phone in working order, then to go on with my ordinary work until a call either came over the 'phone or by a running messenger. I had plenty of practice. Sometimes I would go for a week without getting a serious call, then they would come two or more together. I had to enter all the injuries in a report book, and if they were badly injured had to get them home, and often to hospital. I think it was about this time that I really began to realise what the accident rate was in the mines.

These men came from an area of about eighteen miles in either direction, and taking them home in the early morning was a problem, especially when it was impossible to warn the wives of our coming.

I remember being called to an injured man one Saturday morning—the Saturday of the August holidays. He was a friend of mine, a droll type, and was fond of telling me about his great ambition to have a small-holding. He was always talking about the breeds of fowls and pigs and the cultivation of different crops. Whenever an agricultural show was held near enough he used to take his wife and the four children to it. They would spend a glorious day—and I would hear about it for weeks after. He was going to take his family to a show in the afternoon of the day when this accident happened. He was on his feet when I got there, leaning on a tram. A lump of coal had caught him against this tram, taking strips of skin off his side and otherwise injuring him. He had refused to let them touch him until I arrived. His mate—a middle-aged man—was paralysed with fright because of the nearness of the escape. Very gently I examined the injured man, and found he had a fractured collar-bone and four broken ribs. He seemed relieved after I had bandaged him up, and confided to me that he felt better and believed he would be able to take " the gang to the show ".

We got him outside and started for his home. He was not a lot of trouble, because he could walk and did not make any fuss. The ambulance-driver had been called from bed, and when he opened the garage door in a hurry he smashed three of his fingers. That made another patient for me, and he drove us all the way to Merthyr with one hand.

It was about two o'clock in the morning when we started, and the countryside was asleep. I noticed that the moon looked like an enormous orange up in the sky. The patient's name was Will Evans, and he kept on telling me about what he expected to see at the show in the afternoon, while I had realised that for the first time I was going to visit that place from where my light on the sky had shone. There were no glares of great fires brightening the summer sky that night. They belonged to the prosperous days that were gone. We had to drive slowly along when we got into the town, because there were nasty holes in the streets.

Will lived in a long street of old grey houses, and we stopped at the end and decided to walk down the rest, so that we should not alarm his people more than we could help. As we clattered over the stones of the sidewalk I felt the heat of the night coming from the crowded houses, and although most of them had their upstairs windows wide open, the heat must have been unbearable up in the bedrooms. We could hear the snores of some and sense the heat of human bodies as they slept. Most of them were sleeping away one more hopeless night after another hopeless day that formed part of a succession of such days without work or prospect. They could not have been sleeping comfortably, for many heads came out over the windows to see who walked with heavy boots in that early morning.

We halted outside Will's house and knocked lightly. The upstairs was not high, and we could hear voices speaking, then someone called, "Who's there?"

We always tried to get the man we were taking home to answer, even if he was on a stretcher, for this eased the shock to the relatives. If he could speak he was not so seriously injured, they would reason; for most of them realised at once what that knock in the night meant.

"It's me," Will answered.

"Will, bach, what's the matter?" We could hear the fright in the pitch of his wife's voice.

"Matter? What d'you think is the matter? There's no blinkin' work, that's all." Will's reply expressed his disgust at the wasted journey.

When his wife came to open the door she was reassured by seeing him standing there alone, and we did not show ourselves until he had soothed her by the information that he had "only had a bit of a clout, like". The clothes for the children to wear on their outing that day were folded over the fire-guard and I could hear their repeated inquiries of "Daddy? What's the matter, daddy?" while we helped him to wash. Afterwards, he insisted on his wife fetching a home-made incubator of his so that he could display it. I never asked if he went to the show, but he was quite determined that he was going.

I remember him telling me about another accident, when the doctor had said that a piece of stone had gone three inches into his skull, and if it had gone four inches farther it would have touched his brain. He used to say that sledging was easy work, because if you lift the sledge up it will fall down itself, and that he had an easy job—he only had to pull stones down, then throw them into the tram.

I went to Merthyr several times with injured men. One, an oldish man for a repairer, had a stone drop on him while he was on a staging. It knocked him off the staging and rolled on top of him a second time. Amongst other injuries, he had both legs broken—

one in three places. A blizzard was raging when we got him outside, on the "journey" of empty trams. We had a job to find his pipe, and he refused to be taken on without it. Everywhere was a sheet of ice, and the wind was terrific. It took quite half an hour to start the engine of the ambulance, and while we waited in that bitter weather—it was about four o'clock in the morning—we knocked some people up and asked if we could shelter the injured man in their passage. They were down at once, and so were the neighbours, and a cup of tea was ready in an incredibly short while. Someone came along with a drop of brandy, too, which I am convinced saved the life of that man.

Then, when we did start the journey along the ice-covered roads, we had to travel dead slow. There were several hills, and we skidded sideways time after time, and were stuck at the bottom of one for a long while.

We could not get very near to the house, and two of us had to carry that burdened stretcher over the slippery roadway, with the wind bashing the hail into our faces. While I washed him in the kitchen, I could hear the moans of his daughter who was very seriously ill in the bed in the parlour. I had to bend down in that kitchen, for it was too low even for my normal height.

The men came from such a large and scattered area, and often when they were unconscious or dead it was a difficult job to find their homes, because they were unable to direct us. We tried to take someone with us who knew the place, but that was not always possible. Sometimes we passed back and forth for a time before we could find the right house even when we had the address.

It needs a lot of tact to explain to a sleepy someone whom you have knocked up that you have a badly injured man and you are not sure which is the house. More than once I have had to take a man who was past our help to his distant home, and the first warning his family has had is our knocking at the door in the darkness. A man never forgets one of these occurrences.

About this time I became attached to the Military Hospital Reserve and went to Aldershot for training. The additional knowledge gained was one incentive, and the fact that we got expenses paid and some pay was another, for it offered about the only chance of a holiday that we could have.

Three of us went together, and I was in charge of the party and the travelling-warrant. The clerk who made out this warrant was in a hurry, and his handwriting was never good. Apparently we were going from a straight line to one that had two shakes in it, and the number travelling was denoted by a mark that looked like a semibreve that had sunstroke. His pen had quivered a few times over the date, but it made no difference, for none of us could make out what was on this warrant. The first ticket-collector was quite cheeky to us: he seemed to think that we were having a bit of fun with a busy man, and we determined to get our own back. At every changing-place after that we hurried out and placed our cases and ourselves right in front of the ticket-collector at the gates; then I handed over the warrant and we waited silently with our eyes on the ticket-collector—expectantly. We were not once disappointed—we saw a succession of the finest studies of the bewildered expression that any man could desire. While the crowd clamoured behind us and the collector stared, we waited comfortably.

When at last he gave it up and asked us, "What's on this?" I replied, "I don't know, one of your booking clerks wrote it." The usual solution was to ask us where we wanted to go, and pass us through gladly. On one long stretch, when we had three inspections on the train, that warrant served as a variety turn for the carriage—we were sorry that an inspector didn't come in at each station. One came in, looked at the ten people who were grinning expectantly, gave a long, startled look at that warrant, blushed, and hurried away without asking any of the others for their tickets. When I got back home I asked the clerk where that pass was for and how many it covered. He must have forgotten writing it, for he could not tell me.

We had been idle for some days when we went away one year, and as we had some dole money to draw, it was necessary for us to fill up a form for some relative to draw it during our absence. We went together to the Exchange to get the forms, and as we were a bit dubious, it was agreed that I should fill my form up in the presence of the manager, so that he should see it was all right, then the others could do the same with theirs when they got home. The manager instructed me, and read it after it was completed. He told us that if they all filled theirs up like that, there could be no trouble about getting the money. While we were away the others had the money paid to their nominees without any trouble; but mine was refused, and a great fuss was made of the idea of one of the experts that there were three different handwritings on the form I had filled. When I returned and reminded the manager that he had watched me write every word myself and pronounced it correct, I was paid the money in a very subdued silence.

I had been keeping a bottle of lysol in the house for disinfecting purposes, but had made very sure that it was on a high shelf. Whiie I was away my wife took it down so that she could use some, and while she was busy for a minute, Peter came investigating and found it. When my wife heard him cry out, he was sitting near the sink with the cork in his hand and the empty bottle near him. He was crying and pointing to his mouth. She gave an alarm, and in a very few minutes a crowd of neighbours were there. Shortly a doctor and a nurse, as well as an A.A. man who fancied himself as an expert ambulance man arrived. Peter seemed quite happy over things, and although they gave him treatment, it did not appear to have been needed, and we saw no ill effects afterwards. I believe he emptied the lysol down the drain after touching his tongue with the cork and finding it did not taste nice.

Very soon after this alarm we had a real illness to worry over. Peter had toddled down to the park and got on the children's chute. The park then, apart from the children's corner, was an ash-covered waste. He told us that some big girl had stepped on his stomach and he felt pain. It became worse day after day, until his body started to go black. The doctor called twice daily, and finally I had to rush with him to hospital one night instead of going to work. I did not have time to change the clothes I had put on ready to go to the colliery. We had thirteen miles to go to hospital, and I sat alongside that tiny figure on the big stretcher, while all the way he implored me, " Don't let them hurt me, will you, dad? "

He had peritonitis, and they operated about three o'clock that morning. My wife was very ill when I got home at eleven o'clock next day. At one o'clock I had a telegram calling me back to hospital because he was worse. He was unconscious for nearly a week, most of which time I spent in the corridor or lying in one of the casualty wards listening to the voices of the children in the ward calling for their fathers or mothers and imagining that I could hear his voice.

I used to pass the time watching the lift go up and down in the corridor and thinking that it resembled a pit cage in size, only it was so much cleaner. I watched the casualty entrance, and was not far away when any late accident was brought in. My wife got better to some extent, and joined me in the long waiting. I think he had been unconscious for six days and nights—and we had been warned that we could hardly expect him to recover—when I sensed that a change was coming over that little white face that we watched under the shaded light.

Suddenly he opened his eyes and called, "Is it time to come down, daddy?" From that minute he got stronger, but our worry was not over, for we were not allowed to see him after that for fear of upsetting him. On visiting days we wandered around the grounds trying to get near enough to see into the children's ward, and wondering if he had been moved out on to the verandah, and if a little white arm we could see being waved to one of the other cots could be his. After thirteen weeks we insisted on bringing him home before they wanted to release him. He had altered so—we could not recognise him at the first glance. He had the welcome home that a hero might have had. As only two visitors could see him at a time, his pals made a queue outside the door and quarrelled all day over their turns. The room was half filled with presents. The fish-and-chip man brought him his finest cutlet personally, the sweet-man came down with chocolates, and the grocer brought biscuits. I believe Peter had worried them all with his pranks in the past, but they showed very plainly that they wanted him back—they showed, too, the great sympathy that I have always noticed amongst the people of South Wales.

I was at Aldershot during the Tattoo for my second training period, and was up until one o'clock every morning watching Britain's finest recruiting spectacle. We gossiped with the troops through the day while they rested before the excitement of the night, and in the evening we moved amongst that immense crowd that covered the darkened slopes of Rushmoor Arena. It was all so different from the quiet woods and valleys of our homes. It was very strange, too, to be wearing dusty black clothes one day in the depths of the earth, and the next day to be wearing white overalls and be on duty amongst the spotless precision of that great military hospital. I felt a deal of sympathy for our hospital nurses, for my feet were aching before the end of each day.

On the third visit—by which time I discovered myself to be a sergeant in the St. John's and in the Military Reserve—they put me in the headquarters office. I suppose they thought I looked as little like a clerk as possible, so they made me one. I enjoyed it all right, for it was easier than the ward work, and I had been having my share of that. I remember I made a blunder on the third day. It was something to do with the way I classified a patient who belonged to the artillery—I grouped all the artillery together, or made some trivial distinction, as it seemed to me. I had asked one of the other clerks about this, but must somehow have misunderstood his meaning. When I was called into the chief's room, I was determined that I would not mention the other clerk's statement; after all, they could not punish me very much.

"How did you come to enter this in this way?" I was asked, and the tone was more gentle than I had expected.

I replied that I thought it ought to be entered that way.

"That way! Yes, of course it ought to be entered that way," was the—to me—astonishing reply; "that was the old method of grouping, and it would be to-day if there weren't so many interfering old women looking after the soldier's business."

I was completely bewildered as to what the difference was, but did not say so, and tried to look wise. I was a favourite clerk after that, and was moved up next to the chief clerk.

My neck had got much better of its own accord, and none of the Army doctors had detected that anything was wrong with it. It went rather stiff if I caught a cold, but did not give me much trouble otherwise, nor does it now.

The old type of repairer was passing in the collieries—they had been men from the country districts mostly, who knew one sort of timber from the other and knew where each sort was most suitable. They were careful with their tools, and knew how to keep a razor edge on their hatchet and how to set and sharpen a saw so that sawing was not a strain. They needed to be strong and cool, for they were changing their working-place every shift, but there was danger to be remedied wherever they were sent. In the big seams they often had to lift large posts up on to stagings that were fifteen and twenty feet above the ground, while above their heads the coal-gas was "filling the hole", and under the staging the coal journeys rushed back and forth and the steel ropes swung about and smacked like giant whips.

In their place came a more hurried, but less skilful type. These were the men of the steel age, who placed steel arches in the positions where timber had been standing before, and whose chief tool was a spanner. These arches are comparatively easy to erect if one has sufficient room, but they cannot be sliced or cut shorter, as can timber. We did not like these arches for some time, because they did not warn us with their cracking as timber did; but in time we found that they held better against the squeeze and that they allowed more room on the roadways.

Steel helmets came too, and we were slow to take them into use. One reason was that we had to pay dear for them. I remember the first time one was brought down. The overman brought it and placed it on the rider's head, while he showed the other men that a fairly heavy blow on the head would not hurt a man wearing one of these hats. He demonstrated by knocking the rider on the helmet with a wooden sprag. It was very successful at first, and the overman turned to the others for their appreciation. The rider took the helmet off to see if there was a dent, and at that minute the overman decided to convince one doubter by giving another really hard tap—but he did not notice that the rider had taken off the protection. It took quite a while to get the rider back to consciousness, and he did no more work that day.

We had been losing a lot of time at the colliery,

and a full week was a rare occurrence. It often happened that we would not know that there was no work until we got on the colliery, and we had our journey for nothing and bus fare to pay, as well as bathing all over for nothing; and that latter is no fun when a heavy tub has to be dragged in, and the room becomes covered with dust from the pit clothes.

It often seemed to us that the colliery-owners were in league with the Exchange so that we should lose the time in such a way that no dole would be due. I remember we were working only four turns every week for over six months, and yet not once getting eligible for the dole during that period, because we had to lose three in six to get paid. Often it was very difficult for the company to work the fourth shift, but they did so, and the men worked that shift at a loss, because had they not worked it, they would have qualified for three days' dole. My feelings were not very pleasant when I had to go to work for eight shillings and I would have had fourteen and sixpence—three days' dole—if we had not worked that night, while we knew that the colliery was bound to be idle again before many days—but they would not "link up". All that spring and summer I was working, but was not a penny better off than if I had been on the dole; while the men with big families and who had a shilling a day bus fare to pay were losing money every week by working.

At another period I remember it was arranged for the colliery to borrow trucks and we worked—for half a shift. That half shift prevented us drawing three days' dole, and as we were on the border-line of the compliance period, it meant that we had to lose another six waiting days before we could draw. I earned four and six that day, less bus fare. I would have drawn fourteen and sixpence if we had not worked, and would have kept my waiting time in compliance. As it was I had to lose six waiting days—twenty-nine shillings and the three days' pay—fourteen and sixpence—for the pleasure of going underground and earning four and sixpence. Mine was an average case.

One of my mates—a pretty intelligent type—was injured when some trams ran away and pinned him down. His ankle and one foot were badly crushed, and we got him outside and away for home in Merthyr as fast as we could. He was a wireless enthusiast, and had been bragging to us about the perfect reception which he got from America: a boast which we debated—on principle. On the way home he suddenly remembered that there was an important broadcast from America at three that morning—it was then about twenty to three. He hurried the driver up so that we could get to his house quickly and he "could show this b—— what a real wireless was like". He had no time to pacify his wife when she came to the door. Her frightened, "Jonah, bach, whatever is the matter?" was unheeded. He insisted on holding to my arm and hopping across the room to his set. When he had adjusted it to his liking and the voice across the Atlantic was booming into the room at the early hour of the morning, he allowed himself to remember his pain, and sank back into a chair. Just then a little figure rushed down the stair, seemed to cross the room in one jump, and land in his father's lap in the second movement. A minute later a clean white nightdress was very black indeed.

Later on, after his recovery, it was Jonah who came rushing up to me while I was working on the main. It was then somewhere about five o'clock. I noticed a light coming nearer very quickly, and soon saw it was my friend, and that he was badly frightened. He came close to me, and while I looked at him his mouth worked, but no words issued. Suddenly, in his temper, he hit himself in the mouth with his clenched fist because he could not speak, then dropped on the side and covered his face. I did not need his words to tell me that something had happened. I just unslung the big tin box off the jagged piece of stone where it was hanging, and caught under Jonah's arm to hurry him with me so that I should know where to go. My mate flung the tools into safety and followed. On the way Jonah recovered enough to jerk out something of what he wanted to say, although every word was an effort. He always stammered a little, and the fright had paralysed his speech completely.

When we arrived at the spot, we found that things were not quite so bad as he had imagined. He had seen the stone drop on two men, and had thought them smashed up, but a tram had eased the weight, and been flattened itself by so doing. Both men were under it, but one was rescued with just a few cuts and bruises and a big lump on the side of his head. The other had an arm and a leg broken and his back was gashed rather deeply. He lived fifteen miles distant, and when we got him home we found that his wife was away, so he was doing for himself. We passed the house and got wedged on a sharp corner when we tried to turn back, so that we had to lift the ambulance car back round, as we would have a tram that was off the road. Then we had a job to find the exact stone under which the key was hidden, and lighting the fire was a problem that was solved when we found several candles. Neighbours came in soon after and helped. When the doctor arrived, the patient wanted a flagon of beer—in fact, he wanted two, one to be shared between me and the man who was with me, and the other for himself. The doctor appeared to be a personal friend of our patient, and he consented. While two young women prepared breakfast, another went somewhere for the beer, and we washed the injured man all over. The doctor had approved my splinting, and said we had best take him on to hospital. When we had finished washing the man we gave him the flagon, but we found that the doctor had eaten a good breakfast while we were busy, and had washed it down with our flagon of beer. Four young nurses were waiting for him at the hospital. They had the material ready for washing a coal-black man, but when they saw him, there was a chorus of delighted feminine voices, "Why—he's clean."

We were supposed to rip a stretch of roof that had sagged one night. It had parted from the upper roof and left a crack there. There was a little gas high up all around, but thirty yards lower than this place there was a body of gas from a blower. It was not safe to fire at all, and in any case a hole should have been bored to confine the flash of the powder. To save time we were ordered to place the charge in the crack and clay on top of it. I went fifteen yards farther down to watch that no one came up. When the shot was fired the flash ran down along the place where the top had given and caught in the gas that was in a hole right opposite where I stood. I saw the flame fill out and pass me, and my eyebrows were singed.

The wall of fire rushed down the heading with a noise like the roll of thunder, and the instant thought flashed through my senses, " This is good-bye ". Somewhere between where I stood and that blower this rushing flame met a small current of fresh air, and the flame died away. Had it rushed along for another ten yards, nothing could have stopped that colliery blowing up. But it died down, and only three men out of the four hundred working there knew that the tongue of hot death had licked out at them that night. The fireman spent some hours beating the smell of singeing from that heading.

I once read a statement from a mines inspector in which he stated that they would give every attention to any complaint that was sent to them and the informant need have no fear that his name would be disclosed. If that was true, I wonder what caused the following thing to happen?

I have a friend who was—before he was dismissed because of this happening—a committee man for one of the seams in a local colliery. He had complained to one of the firemen about the ambulance facilities there and the neglected state of the airways. He had not gone past the fireman with these complaints, but had several times quarrelled with this official about the items mentioned. One day this friend of mine was summoned to the colliery office and asked what he meant by sending a letter to Whitehall complaining about the airways and the ambulance arrangements. He spoke what was true when he denied ever writing or sending this letter. He was not believed, but was told that his name was signed to it. He was not allowed to see the letter, and never saw it, but a comparison of his writing proved that he could not have written it. The manager and the inspector of mines have agreed that the writing in no way resembled his, although it was signed with his name. He is almost sure that he knows the one who forged the letter, and has a sample of that man's writing to prove it, but he is not allowed to see the original letter. He is still without work, and has spent all the intervening period of more than four years trying to get justice—so far without result.

CHAPTER TWELVE

Each year I seem to be more concerned with the problems of our living. I have discovered that there is much more in life than eating, sleeping, and working, although it does sometimes seem that those who have no other interests are the most happy.

I have read something like four books a week for a couple of years now, and have gone a long way from the Redskin and cowboy thrillers of my youth. My little library has been gained by sacrifice—by going without a visit to the pictures or to a football match and paying two shillings for one of the cheap editions of some great book. I made the oak book-case in the woodwork class, and now *Crime and Punishment*, *Thus Spake Zarathustra*, *Madame Bovary*, *Mother*, *Wild Wales*, *Lavengro*, *Jude the Obscure*, *Miner*, *Tales by Maupassant*, *Plays and Stories of Tchekhov*, *Resurrection*, and many others are at my elbow waiting to console me when I am lonely.

I have been lucky, too, in having a friend who is a Librarian in a small Welfare Library. The term book-lover has a new significance for me when I watch him handle a good book. Every spot on it is an insult; a torn page is a criminal act. Often when I visit him and sit down until the crowd has been served, he looks at me with that quizzical smile that I know so well. Afterwards he goes to his cupboard and from it he brings some volume or other, stroking it gently as he shows the title-page. "Kept it for you," he says; "it's a good one."

And I know that it is indeed a good one. He spends his working days at a pit bottom, but every evening he has the company of his beloved books—he is a better husband and father because of them.

Then, again, when one starts to think it must surely be realised that every passing week makes one older, and although old age seems a long way off, there is no avoiding the knowledge that a working man must depend on his health and strength to live, so that, if he is wise, he must work every day in such a way that he will be able to work again the next day, the next week, and the next year. There is no place in our industry for the ill or injured, no matter how that incapacity may be caused. A working man is only of any value as long as he can do a hard day's work; when he cannot keep up the pace, he is an encumbrance.

I sympathise with the older men, and watch their struggle to keep up. I listen to the labour of their dust-clogged chests when they climb the drift to go out. They climb a few steps, pause to regain their breath and watch the younger ones hurry past. Down that coal-drift rushes a current of air that is forced and always ice-cold. This meets the sweating men as they come up and chills them to their insides. It tells on chests that are already weakened by clogging dust and the rush of work.

I watch how the few men who are old come to work; how weary they look; how their faces seem almost as grey as their hair; how desperate they are that the officials shall not think they are slower at the work than the younger men. I can read the question that is always in their minds: "How many more times can I obey the call of that hooter? And after I have failed, what then?" It has been pathetic to note the interest they have taken in the pension scheme which was suggested—and abandoned. It was impossible to go on with it because the relieving officers would have deducted that amount of pension from the parish relief, and the leaders knew that the majority of miners over sixty-five had to live on parish relief. What a reward for a lifetime of hard and dangerous toil!

We have made it a rule that if we have anything in our house, no beggar shall ask for food in vain. Last night an old man staggered past here and went about a hundred yards farther along to where a tree shelters a low wall. He stayed there—because he could go no farther. He took off his boots and lay down under that wall. He was there for hours, but did not ask anyone who passed for help. Long after darkness had come I noticed he was still lying there, so I went along to ask if he was ill? He shook his head; then I asked if he wanted food? "Food? Food?" he said, as would one who recalls a thing he had forgotten. While he ate some food and drank some warm tea, I tried to persuade him to go on a little way to where he would find shelter in a barn of a farmer I know. He spoke in a whisper, but was insistent on his need to get work, or else he would have to go "in". He was lying there when I went to work, but was gone when I came home in the morning. I wonder how many more days he evaded the inevitable?

Our treatment of the old people seems very harsh unless they have a son or daughter to shield them, and even those often have not sufficient room for their own needs and cannot properly spare the little that the old ones take to eat. It cannot be conducive to happy eating to know that each mouthful means less for one's grandchildren, and the old chair is not comfortable in a room that has to act as kitchen, dining-room, wash-house, play-room, and dressing-room.

Early last Saturday morning three young men stopped outside this house. After a deal of hesitation one came to our door and asked quietly if we would give them a spot of warm tea. He was very shy, and when we questioned him, he said they could manage for food. I had eaten all the bread for my breakfast after coming from work, so my wife cut them three large slices of apple cake to go with the tea. I saw them put down the "food" they had carried, and by accident or intention this small parcel was left near the stones on which they had sat to eat their breakfast. They meant to walk to London, in the hopes of getting work, and after they had gone I went over to open that parcel. It was a Lyle's sugar packet, and their "food" was inside—one large crust of bread that was terribly hard and had been dusted with pepper as a substitute for butter. Must my boy go through that?

If a man is sensitive and thinks about things, he must surely get to hate the injustice of it all. I feel I hate the continual slavery and dust; the poor clothes and bare living; the need for decent men to beg their bread; the huge van that comes around every Friday and disgorges four beefy ex-policemen, who rush into a house and come out with the furniture of some miner, whilst he stands whitefaced on the side, with his children crying and asking what is the matter; the eviction from his home of some miner who has opened his mouth too wide or refused to be robbed of his wages when they were due.

I think often that if this is all that our two thousand

years of Christianity has accomplished, then the sooner we admit it has failed, the better; but I believe in my heart that it would not have failed if we had more sincere advocates, men who practise what they so loudly preach. I have seen so many men who use religion as a cloak that I have become suspicious of the religionists—though not of the religion.

This world is so full of good things; there is plenty for all. I have experienced some of these good things when I have ignored the "Trespassers" warnings and walked against the wind over the mountain, then sat near a mountain waterfall whilst a primrose took its first glance at the sun.

Then there is music, and most of it was created for us by men who starved as they wrote, yet their tunes live for ever. I was not content to listen, I wanted to play as well, and now I can sit on the next seat and keep note-perfect with an expert violinist. We have arranged our own tunes; we have made our own violins. We saved up to buy a piece of Italian pine for the front and a piece of sycamore for the back; we shaved away slowly at the wood, then bent the sides and adjusted the fittings. Late one night we ventured into an empty concert hall so that we could test the carrying-power and the tone we had given to these pieces of wood, then we looked long at the instrument and at one another, for it had a grand tone, flutey and crisp, as good as we had ever desired, for it is very true that violins are like human beings: no two are alike.

I saved a shilling a week all through one winter, then bought a ticket for a week at the National Eistedfodd of Wales. I do not ever expect to have more pleasure in one week. One of the national papers stated that our area was remarkably prosperous, so our colliery obliged by closing all that August week. I did not mind. I made a parcel of food and tramped the seven miles every morning. The events of each day and the glamour of the evening concerts sent me along that homeward walk very happy each night. The weeks of sacrifice were repaid after I had been near Conchita Supervia and heard her sing. She is the only great artiste I have yet seen.

I have waited to play a solo whilst the air seemed full of insane thoughts because over a thousand mental patients were seated in the hall and a man in the blue dress of an inmate was at my side waiting to press the button that would lift the curtain. I shall never forget the scream that came from one of the wards and tore through that concert hall, nor the whitening face of my lady accompanist. I have watched five hundred sightless children listening to my playing, and have stood in the darkness of their dormitory whilst some of them read with their fingers. I have played to old men and women, and the younger homeless men and women at the workhouses, and every time I have wondered why we do not remove the things that cause this misery; why we do not give the care and attention before the illness or trouble develops, instead of afterwards.

When I played at one workhouse I had three fingers cut with a stone, and they had not healed. I could not feel the strings properly if plaster was over the cuts, so I decided to go on as I was, although the strings opened the cuts afresh. By the end of my playing my fingers were bleeding, so I went around to ask the porter to bandage them. He was writing his report of the last entry for that night, so I waited for him to finish. My hand must have lain for awhile across his desk, because I noticed, whilst he was bandaging, that some drops of blood had fallen on to a slip of paper he had left there with some coppers placed on it.

"Never mind," he said; "it'll not be much trouble to make another note out."

I was silent for awhile, for a reason that I do not think he guessed. I was considering the story of that slip. A bit of paper, some drops of blood, some coppers.

"Frederick Read—total 2½*d.*"

One man's worldly goods.

CHAPTER THIRTEEN

I HAVE had a shock to-day. Of course, we expect shocks on Friday, because it is our pay-day, but this one was not the usual " day-short " or " extra-deduction " sort of surprise; it was the reappearance of one of my old mates—Billy Ward.

I was dressed and ready by one o'clock on that Friday because I had to draw my pay for the previous week's work. That meant that I had to sacrifice some of the broken sleep that I manage to get during the day, and probably accounts for the bad humour that I usually have when I come downstairs. I feel very grumpy, and have little to say to anyone.

The colliery where I work is about four miles from this village, so I have quite a journey to get my money. I could travel by train or motor bus, but usually I prefer to walk on a Friday, so that I shall be sure of being by myself.

It is a grand day, and this valley is beautiful, with the glitter of the slow-moving river down the centre and the several shades of green among the trees that shade the mountain on each side. The sun is warm; some pigeons wheel above the village, and their white undersides show like puffs of steam; the polish shows on the laurel leaves when they move slowly with the warm wind; and the steam that rises from the turf has the smell of rain that has fallen to good purpose. I draw in the beauty of it all, slowly and gratefully, for I feel the wonder of such a day right into my bones. I am not often out in the fresh air, and so I enjoy every moment of it; yet I cannot help that feeling of regret that my mates should be shut away from it all.

Underneath the mountain that is over to my right more than a thousand of my mates are shut away from the sight of this day. They are swallowing dust with each gasping breath; they are knocking pieces out of their hands or bodies because of the feeble light; they are sweating their insides away so that they, and those dear to them, may live.

My home is on the outskirts of the village, and I am soon on the main road. There I walk quickly, so that I can get to the pay-office before the crowd arrives. The road is wide, and a lot of traffic travels along it. The many cars that pass throw out a spray of small stones and grit towards me, for they have been tarring recently. When I have walked over two miles I pass under a bridge that carries the coal-trams from one of the collieries. Just past this bridge, alongside the road, are the screens. The trams are tipped on the moving riddles of these screens, then the coal is sifted until the largest lumps are carried over to the outside row of trucks. There are six rows of trucks; each row is for a different size of coal, graded from the very small to the large lumps; but the colliers are not paid anything for the three smaller sizes. A large pointer on a clock-face shows the amount of small coal that is in each tram, and a weigher books it down so that it can be deducted from the total weight. Whilst these screens are working, a thick haze of dust covers all around. It is difficult to see the way along the main road, and one feels that evening has come suddenly.

The car-drivers rush through this cloud of coal-dust as quickly as they dare, but I have often thought that they should be compelled to stop and made to get out to stand for some time in that dust, so that they could taste it and enjoy a few sneezes. Some lumps of coal could be dropped from that bridge on to them, too, so that they might realise how hard they can hit. These people might then value the men more who get the coal, or, if they had to be with me this afternoon, they might discover some of the misery that is caused by that black mineral which comforts them with its heat.

About two o'clock I arrive at the next village. Here another colliery screen is at work. It shakes the dust into the lower atmosphere, and the sets of coal-trams rattle down towards it between the rows of houses. The dust shaken from the loaded trams covers the windows of the houses, but sometimes a lump rolls off into one of the gardens as some little compensation. The soil in these gardens is coal-dust; cabbages are grey in colour, and are loaded with dust between their leaves; and everywhere one puts a finger, a print is left in that dust. The higher atmosphere is no clearer, because a tall stack gives out a continual smudge of black smoke, which soon releases more soot to fall on the houses. I have heard it stated that nearly twenty per cent. of all the air breathed in this village contains the dust that causes silicosis. Children in the school yard breathe that dust; people in bed inhale it as they sleep; the posters outside the newsagents' are covered with soot, and a dress or collar is not clean for ten minutes.

The colliery office is near the main road. I bend to the window and shout my name and number. When the slip is handed to me I sign it, then get a large envelope in return. The size of that envelope is not a good guide to what it contains. Before I move away I make sure that the pay is right, for no mistake will be rectified after the man has gone from that window.

I shake out my money—two pound-notes, two half-crowns, three shillings and sixpence; two pounds eight and sixpence. That's all correct, so I move past the doorway. A bag of bolts is on one side and a steel rope coiled on the other. There must have been a wall-covering on that passage once—green, I should imagine—but now it is all grey-black, with a thicker coating of coal-dust where the colliers have leaned their bodies against it.

In the room on the left about twenty men are seated on plank forms waiting for the compensation doctors to come and examine them. The signs of injury are plain on most of them, for several have their arms slung, and four are on crutches. It resembles the dressing-station after a battle. Across the passage is another group waiting in a line for the clerk to pay them some compensation. He counts some money out, and calls each man forward to sign for his payment. They will receive not more than thirty shillings, usually less, for a week's compensation, but they are easier in their minds than those in the opposite room, because their claims have been admitted.

Soon afterwards I move away, and walk along the main street. Here, again, the shops are in most cases the front rooms of houses. About fifty yards along this street is a public-house. Somehow I crave for a stimulant to-day. My nerves are upset, and I know there will not be many men inside at this time. I am a poor drinker; sometimes I do not enter a public-house for six months at a stretch, and even then I usually hurry over my drink and get outside as quickly as I can. I do not like the atmosphere of these places, and often wish that the beer-garden method was in favour in this country. About here the drinkers talk so much of their work; probably because they have no other interests. They argue about trams they have filled or driven; about timber they have notched and placed in position; about this yardage and that tonnage, whilst I have always been determined that I will forget the drudgery once the shift is completed. To-day, however, I feel so shaky that I must have a drink, so I push the door back, order a small Guinness, then enter a room on the left. This is a large room, probably the most cheerful in that village. It is clean because no man in collier's clothes is allowed in there; they have their rooms at the back.

The four men who are playing bagatelle are not acquaintances of mine, neither are the two who watch them. I sip my drink slowly. When it is finished I order another, then sit with it before me because two drinks are my limit—I never take more. Whilst I am watching the froth sink, another man comes in, notices me, and comes to sit alongside. He is about my age, but wears spectacles. He has a good trilby on his head, and the crease on his grey trousers is exact. We have worked together and know one another's habits, so when I tell him it is my second drink, he does not bother me to drink again.

I am looking through the window above the

bagatelle table when I hear my companion whistle to himself. I turn to see what has surprised him.

Another man has come in—with the silence of a ghost. His features, too, are ghost-like in their pallor. His skin stretches tightly across his cheek-bones, and his eyes are like two large drops of water in holes that have shrivelled into his face. This newcomer is closing the door when I look toward him, so I can see the trembling of every muscle when he moves. He is hardly strong enough to close the door, then when he succeeds, he sinks into the nearest chair, and his sigh sounds clearly across that quiet room. The barman brings him half-a-pint of beer, takes the coppers from that trembling hand, hesitates near him as if he wishes to ask this customer something, but finally decides to say nothing and walks back to the bar.

The drink is not used. I doubt if the trembling hand could lift it or that slobbering mouth contain it. The man sits alongside the table and trembles, with his cap going back and forth with each shake of his head. He is like a man who is very old, but there is something about his figure that refutes this idea of age.

One of the bagatelle players makes a stroke, then looks around for our appreciation. He notices the newcomer for the first time; the words of banter are choked in his throat. He stares for some seconds, then whispers to his neighbour, and I hear the words he uses:—

"God in heaven, Jack! What's the matter with that chap, eh? He looks devilish bad, that he do. Didn't ought to be about in that state. Took bad in the street, I s'pose, and thought as a drink would liven him up, I expect."

"Where've you been living lately?" his mate asks. "Don't you know a chap as have got silicosis when you sees him?"

"Silicosis? Phew!" the other gasped. "His face looks like a dead man's face, that it do."

"Won't be long first," his mate replied. "I've seen enough of it to know as he can't last long."

I glanced at the man near the door, then quickly away again. He had not touched that drink; his head was still nodding continually. When I look again at the window I feel my companion touching me on the arm.

"I think that chap signalled to you when you looked that way," he said. "You didn't notice him. D'you know him?"

I look again, for I had not seen any signal.

The newcomer notices me as I look, so he tries to smile at me. He does manage a queer kind of grin, but that alters his expression, and in those altered features something that is familiar touches my memory. I look at him steadily—and wonder. It isn't?—no, it can't be?—and yet, somehow he does resemble. Then he gives me that wistful travesty of a smile once again, and I am sure it is.

I move across that room quickly, so that I shall sit next that trembling figure. "Dai?" I question, then I check myself, because I cannot let him see that he is so altered that I did not recognise him; yet I am still not quite convinced that this shadow of a man can be the well-built and healthy friend I knew less than two years ago. Then he nods—just a more emphatic shake between his constant trembling—to confirm the name. I do not know what I can say next. I am not so foolish as to ask a dying man if he feels ill.

" Have a drink," I suggest, because it is the only thing I can think of at that moment. " Put that one inside you, and I'll get you another. It'll buck you up a bit."

" No, thanks." He mumbles the refusal, and I can see his chest lifting with every breath. " I had to get this one—as—as excuse, see." He paused to gain more strength before he continued, " Because I wanted to sell—some of these—tickets."

Slowly, by being very patient, I learn about his trouble. It is more than twelve months since he staggered home after his last working shift and had to realise that his usefulness was over. The stone-dust had got inside his lungs, then every respiration had damaged and torn the delicate lining of the chest—as if rough stones were being rubbed inside a silken pocket handkerchief. This dust accumulated in his breathing-organs, then closed together like cement. When the lungs were torn they were no longer air-tight. Every day his breathing became more difficult; soon it would be impossible. He is a long time telling me about this, and he goes through the details minutely, as would a man who has studied his own disease.

Dai was an intelligent type, and he must know what has happened to so many others, and that his time must be short. I do not attempt to hurry him, and I listen to the record of his visits to the doctors and their methods. As he talks I recall the others that I know who are in a similar state. There are two who are not yet so bad but that they can mount the steps to the Welfare Library if someone helps them; that other who struggles early to evening chapel and totters behind the door to hide his clothes, for he has started listening to a sermon for the first time when he realised that he would not live many more weeks, and that he would never see his thirtieth birthday; and that other of whom they said that the doctors blunted a chisel on his lungs when they held the post-mortem.

Dai is gasping to me that the Federation is fighting his case, and that meanwhile he is living on " the parish ". He is hopeful that his case will be won, for he will then receive about twenty-five shillings a week regularly. I am not so sure about that winning, and I am fairly confident that he will not be alive to see the winning, anyway. I know that these silicosis cases are difficult ones to fight, because the men who are the most frequent sufferers are the workmen who open out the hard-headings through the solid rock to get at the coal.

These headings are completed when the coal is reached, and then the colliers develop the coal-seam. The hard-ground men, as those who drive these hard headings are termed, then move on to another district, or perhaps another colliery, to open more work; so they are moving about every few months, and no company will admit that they contracted the disease whilst at their work. It is sometimes necessary to produce samples of dust from the strata where the men have worked. Often this is practically impossible owing to that part of the mine having closed down, or the continual movement of the workings has hidden it so that it would be a costly matter to expose it again.

Masks are usually provided, but when a man is working to his limit to keep up to the pace set for him, how can he do so if he is handicapped by a mask, for air is nearly always scarce in these new headings? Many men know the danger of this dust-inhalation, but, as they have the choice of this work or none, they choose the course that will shorten their own lives but will give their families food for awhile and a lump sum in a few years—if they are lucky. Of course, every man who goes in these headings hopes that he will be luckier than his fellows, and that he will either avoid the dust, or that it will not affect him. Almost always this hope is vain. Man seems to be very sure that he can beat the mountain, but Nature nearly always wins. I knew six young men who came direct from the farming areas to this part less than ten years ago. They were fine specimens of manhood, and a hard heading contractor gave each of them a few extra shillings to work with his gang as labourers. All have been taken back to that village churchyard except one, who was saved because he had his back injured in an accident. I once complained to a manager that the dust was affecting my throat. He advised me to drink more beer. Earn a little more money so that one shall drink more beer to wash the dust away—it might appeal to some, but not to me.

By slow stages Dai comes to the object of his call. In his more prosperous days he bought a gramophone—" a good 'un it is, with thirty records ". Now he is trying to raffle it so that he can get a little ready money for his wife. He has a bundle of tickets—the sort of tickets one can buy in Woolworth's, that have a perforated centre and a corresponding number on each side. I notice that he has not sold any as yet, because the number one is showing on top; he is shy and does not like asking strangers.

Dai looks at me so hopefully, wondering if I will start this thing going and take that ticket from which the number mocks him. I decide to take a shillingsworth, and to explain things to the wife when I return home. My mate with the glasses takes another shillingsworth after I have explained to him. We approach the others, and the three who are miners buy without any hesitation. The two railwaymen are not so agreeable, but they do buy eventually. That is the thing that I have always noticed in these areas: if a man has a job that is comfortable and secure, he has not the same sympathy for others as the man who is in the midst of it—you will never see a miner refuse help to another who is sick or injured, for it may be his own turn next; but the others who make up the mining communities may live with them for years and yet not take any interest in the problems of those who create the industry on which they live.

Dai cannot rip the tickets for them or sign the names. He holds the tickets out, then they follow the shaking of his hand and do the rest. My mate with the glasses has much more cheek than either Dai or myself. He is a salesman who helps his brother in a market at week-ends. He offers to take Dai around the other rooms and help to sell the tickets. I know he will get rid of a good number, and I want to catch the train home. I follow them to the next room to watch Dai swaying alongside the table whilst my mate sells the tickets. The men there are buying after one glance at Dai; they do not ask what the prize is, they would not claim it in any case.

It must be nearly three o'clock, for already some of the day-men are coming in here after getting their pay. They crowd into the back rooms and shout for a pint to swill the dust from their throats. Most of them wear red mufflers. All have their faces thickly coated with black dust, so that their eyes and teeth contrast with the rest of their features. When they move they leave their trail of coal-dust on the floor and in the air. I know many of them, and they shout their greeting:—

"Hallo, you; going to have one?"

In a small room at the back one of the sub-contractors has placed a row of pay packets on the table and a pint of beer alongside. He is waiting for his "boys" to come from their work. They are really boys, too, for the majority are from sixteen to seventeen years of age. I see scores of them going to work every morning, with their masks hanging from their waistcoats, trousers that seem a deal too wide for their spindly legs, and working boots that they drag over the ground as if the effort to lift them was beyond their strength. Each has a Woodbine drooping between lips that are nearly as white as their cheeks, while their eyes are sullen with the need for sleep. After they get to their work they will be able to straighten up to their full height for about a quarter of an hour, whilst they eat their bit of food. In many cases they are paid less than a pound a week, and many of them—as did the boy who lost both legs last week—have to give five shillings of that in bus fare.

There are seventy-eight thousand boys working in the British mines. Fifty-eight boys under sixteen were killed at colliery work during the first half of 1938, and the records of 1937 show that five thousand, seven hundred and fifty were disabled for more than three days. They may lose a limb or have their features scarred, so that they are handicapped before they have really started to know life.

The old method was for a collier to have charge

of one boy, but nowadays sub-contractors take charge of several boys and make a profit on their labour until they are old enough to claim a man's wage—then they are not wanted. I feel disgusted because these men, who are often critics of the conditions under which they work themselves, should be willing to exploit other people's children for the sake of a few shillings a week. I see these boys stumbling in through the door of the public-house, being handed their packet from the table, and going out again. Sooner or later they will count themselves as big enough to have " one " like the other men; then a small pay packet will be smaller still. If they are to be exploited, surely there is some better place to pay them than in a public-house.

It will not be more than a couple of years before my own boy will be seeking work, but if any sacrifice on my part will prevent it, he shall not go into the mines. I would rather that he did anything else than work under the ground. That is the way most of my mates feel about it. Some years ago fathers took their sons underground with them and showed them what the years of mining work had taught them. They taught them how to timber the roof and part the coal from the top. They felt that they were teaching their children a skilled trade that was of value to the country and would bring them a decent wage in return. All that is changed now, because the conditions are so bad that mining offers nothing but poverty and injury. If a country does not value the lifetime of learning that a miner, or farm labourer, must put into his job, and is prepared to give better wages and conditions to a road-sweeper, can it be surprising that all the workers want to be road-sweepers?

To me the symbol of the miner-boy has been a little fellow who dragged his feet up to the colliery one summer morning last year. I think it was the look on his white face that attracted me. He had cord "yorks" below his knees, just like a real man—it could not have taken many inches of string. His jacket came down to the level of those yorks, and his food-box caused a bulge in his side pocket that was as large as his chest. The poke of his cap—obviously one that he had been wearing to school—covered his left ear. He carried a quart tea-jack which almost touched the ground; he had to lift it higher when he stepped over a rope roller. His lips still held the pout that must have come when he was called before five o'clock that morning, and he dragged himself along as if he hated everybody and everything.

Outside the lamp-window he lowered his tea-jack to the ground so that his hands should be free to hunt for his check. He brought it out like a schoolboy giving a penny for sweets, but the exchange he needed was a lamp, so he stretched up to the shutter, flopped the check inside, then called his number in a shrill voice. I heard a short debate between the lampman and the boy; then the miniature miner turned away without his lamp. I noticed his face; he was smiling now, and the strain had gone from his features. Again he searched his pockets, and this time found an Oxo tin, from which he took a Woodbine. He moved near to me and asked:—

"Gi's a light, butty."

After I had obliged he asked:—

"Want a drop o' water?"

I did not, just then, so he emptied the contents of his jack on the ground explaining, "Shan't want that this shift."

"Why?" I thought I might be able to help him. "Any trouble?"

"Trouble!" He chuckled as he answered. "No fear. There's no work in our district. A fall's stopping us, so I'm hopping it afore the bus goes back."

Then away he went, stretching his legs to their limit, waving his hand in a silent farewell to his friends as they walked past, and blowing out smoke from both sides of his mouth as he hurried along. It was a fine day; he had escaped going underground; and he was the happiest boy for miles. Hundreds of men and boys met him as he travelled downwards. Many of them turned to smile after this blithe youngster—there was no need to ask where he was going.

Crowds of black-faced men were coming from the pay-office when I passed back. Some beggars were holding out their caps to them as they passed. A splendidly built girl was selling flags near the entrance. The falling dust was settling on the cream she had used and was making a grey mask of her features. Her light dress was becoming more dingy every minute. A cheap-jack was offering bargains in working trousers, and an ice-cream man was raising the lid off his cream just long enough to lift a spoonful, then dabbing it down hard again. Several children peered eagerly at the features of the men as they came along, for it is difficult to tell one's own father when he comes home from colliery work; and there was apparently an urgent connection between those children, their pay-bearing fathers, and that ice-cream man.

A queue of colliers were waiting outside a cabin that was no bigger than a chicken-cot to pay their dues to the Miners' Federation. Mine will have to wait for this week, because I am already a shilling short, so I cross the colliery sidings, duck under the couplings of several coal-trucks, and reach the station platform.

On this platform the porter has always a job waiting to be done; he has to be a Canute in a dry way. At frequent intervals every day he starts to sweep up the dust, but he never clears it. There is a thick ridge of dust several inches high which shows where he was interrupted in one of his efforts. When he re-started he began anew at the limit of the platform, and that smaller ridge was not carried to the other, but waits, in a line that is ended by the sloping brush, to be pushed up to join that other soft ridge which is about as high as the top of our boots.

A colliers' train is shunted into one of the side lines whilst the ordinary train passes. There are no cushions in this colliers' train, and the backs of the seats show black designs where the bodies of the miners have rested. The chalk drawings inside several of the compartments show that there are artists amongst these workmen. I can recognise Chamberlain and Claude Hulbert quite easily; the drawings of well-known actresses are neither so well executed nor so discreet. I know that in every little niche of these carriages I should find cigarettes or matches—or pipes—hidden if I were to search, but I am not going to rob my mates, because I know what it means to go eight hours without a smoke, then to hurry down and find the one that has been treasured is stolen.

In the stretch of wide valley which is between the dust-covered area where I work and that other grey area where I live, there is some beautiful country. The train rocks past the rows of loaded coal-trucks, then puffs happily into that green valley between two great mountains. Farmhouses show like spots of white on a dark green board, and a mile away a church spire rises amongst magnificent trees. Between these trees we can see the white-painted sides of a large mansion, and nearer to us a dozen red chimneys show where another mansion is situated.

Even the train seems to groan when it has to stop at the station where I live, yet Friday here is about the gayest day of our week. Soon there will be a little money to spend, so everyone is alert. Most of the streets are of grey stone houses, huddled closely together, with the doors and windows painted the same colour by the men employed by the colliery company. If the woman is house-proud, she polishes the door with one of the furniture polishes; if she is old enough to have become disheartened, she leaves it alone.

This Friday as I passed I noticed that two green-grocers were waiting at the upper end of one of the streets. Their carts were loaded, and the one pony had just discovered that the rival firm was selling nice apples, and was enjoying them unnoticed. A group of tally-men and insurance collectors kept up a desultory argument at the corner, but reserved their energy for the rush that would start soon.

A fish-and-chip cart was waiting to oblige those housewives who had not found the time, or inclination, to prepare the afternoon meal for their men. A young cockle-woman wearing boots that hid most of her shapely calves, and with a plaid shawl pinned across a full bosom, walked down the street with strides that tautened the short skirt she wore. She had a white basket on her right arm, and turned her head as far as the balancing of the tub on her head would allow to inquire of a woman standing in her doorway:—

" Shu da chi hethi? Cockles hethi? "

Then a rattle of steel-shod boots on the roadway showed that the miners were coming. They hurried along, and their groups thinned when one turned into his house, or another stopped to greet a welcoming child. Almost every man had his pay held tight in his left trousers pocket, whilst his right arm was swinging wide to help his walking. Coal-dust rose from their clothes as they moved along, and their gleaming eyes, bright red lips, and white teeth showing from the black faces made me think of nigger minstrels. Each had a tea-jack in one jacket pocket and a food-tin in the other. Some kicked their boots hard against a stone and took off their jackets before they entered the house, so that no more dust than was possible should be carried inside.

As the men arrived, the " chip " man poked his fire and stirred the potatoes in their boiling fat. His pony obeyed the command to " Git up, thar ", and then he disturbed the afternoon with a scream of " All 'ot. Chips. All 'ot."

This call brought several clean-aproned women to their doors. They waited, holding basins in their hands and sniffing at the smell of the chipman's wares—and they hoped their man would arrive with the pay before the chip-cart passed their doors. The collectors suddenly became alert, lifted their cases, wished one another a dubious " Good luck ", and started their rounds.

We had a visitor at our house when I got there. I should have known that it was Billy Ward by that habit of rubbing his finger along his chin, even if I had not seen his face. His face would not have guided me so well, for he was altered very much. I had the affection for Billy that every workman feels for a mate who has shared years at a dangerous job with him. I remember when we were working a lot of overtime and Billy came to work displaying a packet of Player's cigarettes.

"Have a fag?" he asked. "It ain't a Woodbine, either. It won't last for long, I'll bet—these good times when a chap can have a decent smoke."

Obviously the times have not been good lately for Billy; he is smoking Woodbines now. He has a cup of tea with us, and tells us that things are "middling". He still has that habit of starting to laugh when telling you something, then breaking off to chuckle, as if he were finishing the enjoyment on his own. He was always rather sensitive and shy, so that, as I know him so well, I realise that he has something he wants to tell me—alone. Under the pretence of seeing round the old place, we go out.

"Just as dreary as it was, like, ain't it?" Billy looks at the grey streets and sees no change in them.

"Aye, just about the same," I agree. "Let's go up the mountain side. You didn't have mountains like this to climb at that place you went to."

"No." He shakes his head.

Purposely I walk towards the cottage where he used to live. As we walk I recall the times when he used to be waiting for us in his working clothes. Sometimes he had that old cornet at his mouth and we could hear him practising as we climbed upwards. He had been an Army bandsman in his youth, and was a fine player. When we get above the village we pass along a path bordered by stunted trees. I have to push the ferns aside as I walk. We pull a few of them in a heap, then sit on them.

"How's the work going now?" Billy asks.

"It's dragging on," I answer, "and that's about all. We don't lose much time. We don't need to if we're going to live at all. What sort of a show was that where you went?"

"Middling." Billy looks down at the village, seems to be swallowing something hard, then adds: "Hutch is dead. Did you hear?"

"Hutch!" For a minute I cannot believe him. I suddenly remember that Hutch had gone with Billy when he went North to work a machine in a colliery that was installing them. I had the same offer, only I had got work earlier.

"Good Lord!" I continue, "that's a staggerer. Why, he wasn't more than thirty-five, and he was as strong as a horse. Couldn't think of him dying so young."

"Killed he was," Billy muttered, "right by me, and he ought to be alive now."

"Oh!" I answer. "What was it—a fall or what?"

"Aye, a fall it was. A big one."

I let him go on with his story. It is obvious that he has been longing to tell it to someone who would sympathise. As he used to tell me his worries about payments, or children's illnesses, in the past, so now he tells me about this new trouble. I visualise Hutch again. He was broad in build and fastidious about dress. He was fond of a buttonhole and a pipe that smelt of good tobacco. He had a boy's mischievous ways, yet he was a fine man in appearance.

According to Billy, the machine-mining was not a success, and it was soon abandoned. Billy and Hutch were found work as repairers; that had been agreed upon in case the experiment with machines failed. According to Billy, the officials there were slave-drivers; the one who was in charge of their district he classed as " a damned swine ". Knowing Billy so well and his kindly nature, I guessed there was good reason for that remark.

" Oh," I said, " so we haven't got them all down here? "

Billy shook his head slowly. " What's the matter with men, mun," he asked, " that they're willing to treat other men like they do for the sake of a few bob a week more? I couldn't do it myself."

I think of his kindly nature and believe him. He has an under-manager's certificate, but would not be enough of a bully to do the job.

" Give some men a shilling a day more," I said, " and call 'em a boss, and they'll drive their brother to death."

" This bloke would," said Billy. " He was chasing the men all the time. They was starting-men, and if they was under their money and couldn't earn the minimum, they wouldn't get it. If they made a fuss they was sacked, and the manager would see as they was penalised at the Labour Exchange. Then they'd apply for more men at the Exchange, and when them men was sent they'd be treated the same. Me and Hutch had moved our things and was sharing a house. We was in a hell of a sweat all the time as we would get the poke, and we didn't know many people about there. When you been at a place a long time you gets to know where you can get some help, or where there's likely to be work."

" Aye, I know it's like that," I agreed.

" Hutch was saying that he'd flatten the face of that fireman one day," Billy continued, " 'cos you didn't want to hurry Hutch up, he'd slog for all he was worth."

I knew that to be true.

" And this day," Billy continued, " we'd been ripping some top down and hurrying to clear it for the haulier to pass. Our singlets was soaking, and we was thinking it was time to have a bite of food, when the fireman comes running and bawling to us: ' Bring them tools with you, and shape your blasted selves. Come on.'

" A fall it was," Billy explained, " on one of the headings. Nasty it was, too; it had fell up to about eighteen feet, and the top above was like that black pan as you sees about here. You know, it drops without warning. The place was all alive; it was pinching up above and back along the sides, but that fireman wouldn't listen. ' Settling down,' he said it was. And there was six trams of coal the other side of it as he wanted to get out."

" The fall was blocking the rails, was it? " I asked.

" Aye, it was," Billy nodded—" must have been about eight trams, but he reckoned there wasn't more'n five."

" Same old game," I commented.

" Aye, that's it. And we had the wind up. Hutch was afraid too, and he wasn't often frightened. Honest, it wasn't safe for a cat to be there. Hutch climbed up on top of the fall and sounded the upper top; it was like a drum. It was shaking when he touched it. Hutch started to argue, and wanted to put something under it before we started to work, but the fireman said he was going to have that coal out before the end of the shift whatever happened. I said it wasn't safe to work under it, and he started to rave at us. 'Repairers, you calls yourselves,' he said, 'I'll bring you a couple of kids' suckers to play with. Get on at the job or take your tools out.'"

"Wonder Hutch didn't land him one," I said: "he didn't like being roughed about."

"You know how it is," Billy explained; "we'll always get the worst of it, no matter who's fault it is. We asked him if we should put a long post in the middle of the road so as to shelter us a bit, but he shouted, 'You knows damn well that we couldn't get that coal by if a post was put in the middle, and anyway it'll take half an hour to get timber here. For Christ's sake,' he said, 'get on with it, or get to hell out of here.' Well, we had to start, and we was all on edge with little bits a'falling around us all the time. We pitched in for all we was worth to get from under it all as soon as we could. After we had walled all as we could on the sides, he went to fetch a haulier and a tram. He'd been standing by us and shouting at us all the time."

"It upsets me always, when I'm under a bad piece, to have a boss standing near me," I stated.

"I should think it do upset everybody," Billy agreed; "and it was like that with us. We didn't know which was the worst, the fireman or that bad top. We had a tram, and whilst we was filling it, he went off to see as the haulier was back quick. We was about half full, and Hutch asked me to give him the sledge. I give it into his hand, then turned round to chuck a stone into the tram, and as I was doing that I felt the air going by me face, and something clouted me in the back—like a kick it was. Sort of unconscious I must have been for a bit, but I'm sure I heard Hutch squeal and give like a gasp and a long groan. I didn't know what was up, 'cos it was all so dark."

"Lamps knocked out, I suppose," I asked quietly.

"Smashed, they was," he explained. "They found 'em afterwards. I felt like as fire was in front of me eyes, and blood was running down me back. Hit me in the back it had, but not so hard, else I'd ha' been like Hutch was. Was too much small coal in my mouth to shout for a bit, and me eyes was clogged, but as soon as I could I shouted for Hutch, but he didn't answer. Then I knew as it was all over, 'cos Hutch was one who would squeeze through if there was any chance."

"Aye, he was! Poor old Hutch!" I agreed.

"And I felt about me, but only the big stones I could feel, so I started to crawl for help. I felt me way along the rail. Could hardly drag me legs, and I had to go a good way afore I met the haulier a'coming back with the fireman. They was trotting the horse, and they nearly run over me. He sent the haulier back to get help when I told him, and he comes on with me. There was no sign of Hutch. Dozens of trams of muck was on top of him. Big stuff, too."

"What did the fireman say then?" I asked. "Was he upset?"

"He looks at the fall, then he say to me, 'What have you got to say about this?' I says, 'I've got a lot to say about it.' 'If you're wise,' he says, 'you'll say as your mate was scraping a place to put a post, and I'd gone to see about one being brought, when this fell and caught him. Don't forget.' I didn't say much. I was pretty bad meself."

"How long before they got him out?"

"Sure to be nearly an hour. Had to use posts and rails to get the stones up. They was a time before they could find him, and when they did—his head was flat almost, and only a bit of skin was holding his leg."

Billy was biting his lip to help his control, and I had no way of soothing him.

"See," he continued, "I wanted to wash him, because it seemed the last I could do, and he was my butty, but when I saw his face by daylight, it finished me. It fetched me guts up, and I ain't been able to look at food since. And we had to put him in the front room because there was only two rooms down. His wife and the two girls lived in the kitchen with us."

"How are they now?" I asked gently.

Billy did not answer for a minute. He looked down on the village, from which the cries of the hawkers and the calls of children came faintly up to us. A coal-train was moving out from the siding. Hardly a trace of smoke showed amongst the heat haze of that afternoon; it seemed that the loaded trucks were pushing an unwilling engine along the line.

"They was fond of Hutch," Billy stated at last, "and we was all fond of him. I wanted to kick them blasted chaps as came nosing about news; I wanted to smash 'em in the chops. Pretending as they knows what goes on underground, and a kid knows as much as 'em. A miner's kids knows more; Hutch's kids do. 'Unfortunate accident,' they says, and they gives it a couple of lines in a corner. Pah!"

"But they don't know about things, Billy," I tried to console him. "How can they know?"

"Well, why don't they admit they don't know, then, and shut their mouths?" he demanded. "But that manager and that other official as came to see me, they knew all right."

"Came to see you, did they?"

"To show sympathy; but they wanted to see me alone. Told me they had the fireman's version of it, and s'posed as mine would be on the same lines. Pity it happened just as the place was going to be put safe like."

"What did you say to that?"

"Not what I wanted to say. Hutch's two little girls was sitting dazed in our kitchen, and his wife was unconscious upstairs. I hadn't slept or had grub since it happened, and my back was pretty bad. 'Sides, I had to think of my own living. I didn't have much of a job to pretend to be stupid about it all. Acted that way when I went in with the inspector, so I did."

"Was things just like you expected when you went in?"

"Mostly. The tram was flattened. It was that tram as eased the stone off me. It looked devilish rough. Lot of timber about there then, though. Some under the fall as didn't ought to be there."

"Some stood up too?"

"Aye, and I couldn't remember seeing a lot of them when we was there before. 'Course I was a bit dazed, and I feel mostly dazed nowadays, but I could swear as they was put there since the fall.

Smack your cap agen the hatchet marks and they looks days old. You knows that."

"Yes, that's an old one."

"I slept on the couch in the kitchen till the funeral, and Hutch was in the other room. Had to keep the lamp lit 'cos I was frightened to be in the dark, and the weather was hot. It was hellish. I don't know how them women stuck it."

"How did the inquest go?" I asked.

"'Accidental death,' of course. It had to be. They made his wife go there to identify the body. What do they want to torture a woman that way for, eh? They knew well enough as it was Hutch."

"It's one of their rules, I suppose, but it seems a damned silly rule in the cases where I've been," I agreed.

"They said as the place had been examined just a few minutes before the first fall, and the fireman said they was preparing to make it more safe by putting timber there. The solicitors asked a couple of fool questions about colour of stones—as if a stone would go red before it fell. Properly got my back, they did between them."

"You got to allow that they don't know the conditions, but they do try to be fair as far as they knows."

"Some do, and this coroner did. He asked me if I had heard anything said about putting up more timber. I had made up my mind as I wouldn't tell a lie anyway, so I says, 'No.' I hadn't heard a word, and I said I hadn't. Then they asks the fireman agen, and he said as he had mentioned it to Hutch when I was getting the tools. I wasn't ten yards away, and I never heard a word said about timber, but Hutch was dead, and they knowed it."

" What got my back up," Billy continued after an interval of musing, " was that the fireman said as he felt uneasy about that place, and stayed there near us to watch over our safety. He had only gone for a minute when it happened; and they lapped it up like milk. The coroner praised him in the end for his carefulness—his carefulness, mind you! His own skin was about the only thing as he was careful of. They didn't ask him how it was as a place as he had examined and found safe fell a couple of minutes after, or how he left men to work there when he was looking for timber, or how he could have had that coal out if he'd put posts in the middle as he said. Oh no, they took it in like good 'uns, an' looked as wise as a lot of owls."

" You could have told them different." I felt ashamed of myself when I said this to him.

" Aye, and would they have believed me? And what about my job and the family? "

" You would have done good for others, perhaps."

" P'raps it would be. And I wanted to do it, and have a smack at that swine of a fireman. It wasn't being afraid of them as stopped me getting a bit of me own back—no, it wasn't that. I wanted to shout out the truth."

" Why didn't you, then," I suggested. " You might have got support. The Federation would help you."

" S'pose as they would. But what about Hutch's wife and the kids. D'you think as I could do harm to them? "

" Harm? How d'you mean? " I asked.

" You ought to know well, if you was to think for a minute. What about the lump-sum compensation? If they hadn't brought in Accidental Death, his wife wouldn't have had that compensation, and the insurance company would dodge liability if there was any question of neglect about him dying. They had their lawyer at the inquest; so I shut up my mouth and let them say what they wanted. I couldn't let old Hutch down, and yesterday they was arguing as to whether he was worth more than four hundred pound. My God, it's enough to make a man sick."

"When are you thinking of starting again?" I asked. "As soon as your back is right, I expect."

"That's why I came down—to see if there's any chance of a job about here again. I'm staying with my sister-in-law until Sunday. D'you think there's anything going up where you are? I'd do anything—labouring or anything not to go back."

"It's mighty full with us," I said, "and I don't expect there's much chance, but I'll see what I can do. I'll ask the manager to-morrow morning if you like."

"Will you? Tell him I'll do anything as long as it's a job. P'raps a better one will come along later. I know as I won't be many weeks up there now, for that fireman will remember that I hesitated about saying he spoke of timber. He'll have me on the dole as quick as he can, or p'raps the dole will be refused, because he's sure to work up something about insolence, and how can I be so civil as he wants when I knows as Hutch would be alive now if it wasn't for him, and that I might be as dead as my butty is but for a bit of luck—if it is luck? And how can I live in that same house and work in the same place after what have happened?"

" I know what you feel, Billy," I assured him when we started to go downwards again, " and if it is possible for me to help you, I will."

" I knew you would when I came," he answered, and I was grateful for this tribute from my old mate.

CHAPTER FOURTEEN

I AM enjoying those last few minutes before I must change my clothes, for I always leave it until the last minute before I start to don those dusty clothes which carry the hot smell of sweat that is the scent of the mine. A cloth is over the dishes on the dresser to shelter them from the dust, a sack is covering the mat in front of the fire, my pit-clothes are warming in the heat of the coal they have helped to produce. I know exactly how many seconds it will take me to hurry into those clothes, and every minute without them is enjoyable, so I delay to the last.

To-night that constant dread of my work is very heavy on my mind. I would like to go to bed like any ordinary human being, and wake in the morning to go to work that is safe and well-paid—as thousands must be doing; but I cannot stay away from my work for one night, because one lost shift would mean that the bonus shift—which is only paid when a man works a full week—would be lost, and I should sacrifice two turns instead of one; and even that one is more than I can afford.

I expect the news about Hutch has upset me. If it was Billy speaking, he would say, " No, lad. I'm not coming to-night whatever we goes short of. Them bells is a-ringing agen, and when I hears them I stops away from the pit." I have steeled myself to ignore all sorts of fears and warnings, for of what use is it to stay away one shift when the next is exactly the same, and the next? I have been lucky enough to avoid serious accident to myself over all these years, but I cannot check the feeling that some day I shall be just that important second too late when jumping back from a stone which is falling, or will be just a yard too near a tram when the rope snaps—and somehow I feel that to-night is going to be a bad night for me, or some of my mates. Yes, there it is again; it may be one of my mates, and my skill and experience in ambulance work will be needed badly. No, I cannot stay away.

It has started to rain again. I can hear the drops "plopping" into the pool that is outside the back door. The colliery yard will be a stretch of paste made from coal-dust and mud.

The sound of music comes to me from the public-house next door. A wonderful voice is singing—it must be a record being played. I recognise that it is Paul Robeson singing "Old Man River". If I have to run with my clothes in my hand I will not start dressing until he has finished. To do anything else except listen would be to insult one whom I count as one of the greatest men of all time.

You can feel the suffering of the negroes in his voice as he sings:—

> "I'm so weary, and sick of trying;
> I'm tired of living, and scared of dying."

He is singing of a distant river and of slaves, but he might almost be singing of the mountains here and of us. I'm very tired of trying, but I am not so scared of dying; only of dying under the mountain and of being brought out looking like one of the cockroaches that has been crushed by a working-boot. When I must die, let it be in daylight—or in the open air, at least. I am scared, too, about what may happen to the ones I leave behind, and that fear makes me act like a slave; but I have not the value of a slave, because there is no purchase price to pay for my body, and there would be many ready to start at my job to-morrow. Probably we will see some of them waiting at the pit-mouth to-morrow morning, hoping that they will be given a chance to start on the following Monday. The week that is closing may have brought a vacancy in some way. A man may have been seriously injured, and his place may have to be filled; industrial disease may have finished another; some other man may have become too old; or, most unlikely of all, there may be some development of the workings. I must hang about to see the officials—a thing I detest doing—to plead for a chance for Billy; perhaps the fact that he is a stranger to this area may help him, because they can have nothing against him.

The song ceases, and someone who is inside hammers his mug on the table and shouts for "Another pint, please". He has adopted one way of easing his troubles, for a short while. As his knocking stops, the clock begins to strike: One —two—three; it strikes ten o'clock. I know it is ten minutes fast, so I have exactly enough time. At ten o'clock the bus will be waiting on the top of the street to take me to work by eleven o'clock. I can delay it no longer; I start to pull my evening shirt over my head.

At the top of this street is a wide smudge on the stone wall. We have caused that smudge by leaning against the wall in our working clothes so that we can feel the heated stones that carry the warmth of the fire from the kitchen of the top house. I have no need to crouch there this night, for the bus is on time.

The interior of this bus is lighted by one weak lamp, and after the usual banter on entering, we sit back silently, and are carried forward grimly, as if we were going to battle. When the bus bumps over a gutter, a cloud of dust arises from our clothes, and everyone coughs or sneezes. Someone makes a half-hearted complaint about the smell of the tobacco that "Crush" Williams is using. He wants to know, is "Crush" "smoking old linoleum?" Crush saves up a large mouthful, then blows it strongly in the critic's face.

"Got to keep up good steam," Crush explains, "to help this 'ere bus up them hills."

Crush is one of my mates. He is strongly built, an old footballer and boxer, and has earned his nickname by his method of settling any dispute. He does not bother about a long argument, he tries to crush anyone who disagrees with him.

Outside the village we were enveloped in darkness. The woods and mountains showed on either side when the sway of the bus turned the lights from the wide road. At intervals a car rushed towards us from the darkness, glared at us with dazzling eyes until it had passed, then winked a red eye at us as we slid apart. Here and there a lonely house attracted our sight as we rattled along, and I kept up my rule of always having a long look at that especial house—the one with the sycamore trees sheltering it. To-night the moonlight was just strong enough to shadow the walls and the garden hedges.

Crush noticed my interest.

"Like to be there to-night, wouldn't you?" he asked.

"I would that," I agreed.

"Me too," he said. "Reminds me of the old place at home. Bloke ought to have a house by itself and a garden; not the three flagstones as they calls the garden where I hangs out."

The screens were working at full speed to get empty trams ready for the night shift. Dust floated in the air around every electric lamp. We slid in behind two other buses that were unloading, and almost at the same moment a workmen's train pulled into the station near and the men's heavy boots started to rattle across the platform.

There was a little delay in our bus, because it was pay-night, and we all had to settle our debt for the week with the driver. I was about the lowest payer, because I live the nearest, but I owed two shillings. The one who travelled the farthest owed seven shillings for the week, and had started to dress at half-past eight that evening.

Crush walked up with me, and we had to wait awhile near the third window of the lamp-room because they were attending to the afternoon shift, who were giving their lamps in on the other side. Five hundred lamps going in on the one side; lamps that were not so clean and bright as they had been eight hours earlier, then about three hundred clean and bright lamps being handed out from the other side to the night shift. Lights dancing in, dancing out, like a procession of large glow-worms; and a row of lamp-men who handed them in or out without many of the men having to shout the number of their lamp-checks.

Crush was waiting for me near the board that warns of "Danger. Cigarettes or matches not to be taken past this point." He was enjoying a last pipeful before hiding his pipe that is the terror of our bus under an old tram. He has his shoulders thrown back and his eyes looking up at the sky as if he were waiting for a high-punted ball to drop. Six foot tall, and broad with it, he looks a fine figure, even in his rough clothes and with the straps below his knees. He sticks his thumbs under his belt and returns his gaze to earth and to me.

"Ready to get into the spake, guy?" he asks.

I am, so he hides the pipe and we move inwards inside the nine feet of circular arching. I once offended Crush, temporarily, by saying that he would have made a good policeman. He soon forgave me, because he allowed me to say things to him that would probably mean sudden death to anyone else.

The "spake" filled quickly, and everyone hung his lamp over the side where he sat, so that each tram had eight lamps hanging outside it. At ten minutes to eleven the outside rider hurried across, shouted his warning of, "Look out now, boys, she's going"; then drew his file across the signal wires. A bell rang in the nearest engine-house and a red light winked at each ring—three times. The driver lifted the lever from neutral, pulled the steam lever towards him, and the drums started to move, tightening the steel, snake-like rope that was coupled to the last tram. The rider jumped on the coupling as we started to move inwards along the mile and a half of main haulage-way that leads to the districts. We went in like a lighted snake, with our lamps showing the rock edges and the steel arches as we passed. All had their coat collars high, so that the draught, or the water that dropped from the roof, should not get down their necks. The rope sheaves screamed as we passed; the rollers rattled as the rope slid over them; the trams jolted over each uneven rail; and the three hundred men talked of their work, of the football match next day. I think Crush spoke for them all when he sat back against the iron side of the tram and pushed his knees between mine as he remarked:—

"Well, I hopes as half-past six to-morrow morning won't be long, and as I'll have them four aways right this week, anyhow."

We are about the first to reach the bottom of the drift, where the fireman is waiting to test our lamps and tell us of anything unusual that has to be done. Crush is one of the coal-face riders, and he is told of the full trams that are to be moved. He will have to bring empty trams to me also during the shift.

"Won't want the boring machine to-night," the fireman tells me, "there's a fall on the left-hand rise heading. It fell just after the other shift passed. Clear a way to get a horse inside, so as the haulier can go on."

"How much is there?" I ask.

"About six, so they said," the fireman replies.

The man who is to work that shift with me is standing alongside, and we move away together. Crush falls in line, as it is getting narrower, and is now awkward for two men to walk abreast.

"Six trams, that's what he said, wasn't it?" Crush inquires. "That'll mean eight or nine, I'll bet. I knows these officials and their guesses. Good counters, but damned bad reckoners-up, that's what they are."

We have become shadows that carry lamps, and can only judge who is speaking by knowing the voice. Our lamps are heavy, and we change them from one hand to another so that they shall not check the circulation. Crush wears a cap-lamp that is handy for his job, but would be inconvenient if one were working in the same place for awhile. The long avenue of lights that has been following us gets fewer in number after each branching roadway that we pass. The crowd of men has dwindled to a dozen. We have another half-mile to walk.

In some places there is ample height for us to walk upright, and in others we have to stoop almost double. It is this variation in height that causes many knocks on the head. A walking man may duck just a shade too late or a sharp edge of rock or broken piece of timber may cut his head as he stumbles forward. Crush steps away from a piece of timber that is pointing downwards and is about the level of his eye.

"See that bit," he asks me—"just a flick of the head and a step sideways. But allus keep your balance; that's ringcraft, that is."

Nearer to the coal-workings the sides and roof are not so solid as they were back on the mains, where the pressure has ceased because everything has been pressed solid. Soon we begin to feel the heat coming to meet us, then we sense the movement of the newly disturbed ground; the continual creak of weakening timber; a snapping sound as the roof begins to crack above us; and when we are within a hundred yards of the coal-face we hear the roar of falling coal and the sharp staccato cracks as the gas loosens more slips of the coal.

Here is our working-place for this shift. We hang our clothes on a projecting stone out of the reach of the rats, and give a sigh of relief when we have stripped to our singlets. Can you remember a terrifically hot afternoon in summer when the heat-haze prevented you seeing the sun and when the air seemed stifling and pressing you down? It feels as if the heat of all that afternoon had been confined to this underground roadway. Our singlets are sticking to our backs; we sip some of our water gratefully. It is still cold as when we drew it from the tap, but soon it will be lukewarm.

By the time we have released our tools from the locking-bar the haulier has arrived with his horse. He waits behind us whilst we peep cautiously at the fall. Yes, it is more than the estimate—we had expected that—but it looks very ugly as well.

"All alive, it is," my mate mutters; "and it's squeezing over us and back on the sides as well."

It is squeezing all around. Stones that have been walled on the sides are crumbling to pieces because of the pressure; there is a continual crack—crack all about the place as timber breaks or the stones part into smaller divisions. We can see the stone walls—dry walls built of fallen rock—on either side are being pressed forward, and that the rails on which the trams must run are being disturbed by that moving weight.

The haulier must get the other side with his horse, or else the work will be held up, so we reach forward from the shelter of the timber and jump back with a stone in our arms. After we have done this for some minutes and made a shallow lane in the mass of stones, we scrape smaller stones back with a mandrel stretched out to the limit of our arms. We clear away until we know that any further clearance may undermine the other stones and cause them to slide, then we prepare for the haulier to pass.

We show light to the top of the eighteen-feet "hole" that the fall has made, then we listen.

"Seems quiet for a bit," we say.

" Right, then," the haulier replies, " I'll try it. Keep your peepers on it in case anything drops."

He shakes his lamp to give the maximum light, lowers his head and clenches his fists, then he dashes forward like a sprinter. He stumbles over a stone, but does not slacken his speed. When he has rushed about twenty yards and is under comparatively safe roof he pauses and turns round.

" I'm across, anyhow," he tell us. " Send the hoss across after me, will you? I'll show him light."

" Think he'll touch that lag as he passes? " I ask.

" Ought to pass," argues the haulier, " 'cos he's a quiet one and won't kick about."

We study that lag, which is a broken timber support pointing downward. If the horse touches that stick on his way over, he will disturb many more tons of rock which are hanging above, and that touching will probably be the last thing he will do. We cannot move that lag before he passes, or we shall have a larger fall, and no horse inside to help us clear it.

We reach forward again and scrape a couple of small stones away, so that his hoofs shall sink lower and his withers not be so likely to touch that sinister piece of timber. As we encourage the horse forward, the haulier advises us:—

" Let him take his time. He's a fly old horse, he is. He knows the ropes, and I don't want a hoss on my docket."

He is a likeable old horse, and he rubs his nose against me when I let go his bridle after I have led him under the last pair of timber. I am a lover of animals, and I cannot resist the desire to pat him several times on the neck while he stands waiting; for I feel that the old horse may soon be taking his last few steps, and he has not had many pats in his life.

"Come on, laddie, over you go." We get behind and urge him forward.

"Come on, bachgenni, gently does it," the haulier pleads from the other side. He shades his lamp so that it shall light the way without dazzling his horse.

Slowly the horse moves forward, and picks the chosen track as if he were a man. He rolls about when near the lag, and we hold our breath, but he gets steady again and walks across safely. I noticed that there was not an inch to spare over his back when he went under that lag; if he had been harnessed ready, he could not have got past. While the horse is blowing and snorting to show his indifference to bits of fall like that he has just passed, and the haulier is telling him what a "tidy feller" he is and is wondering "where the heck that afternoon haulier left this horse's tack?" the colliers and labourers arrive. There are ten of them altogether, and they pass over singly, and in the hurried manner of the haulier. We keep the only boy—Jacky—until last. He is just over sixteen years of age, alert and very confident in the companionship of men.

All take a special interest in the safe passage of Jacky. Each holds his lamp to show more light; we have a longer period of watching in case anything should fall; all encourage him with: "Over quick, now, Jacky." We all wait poised, in case anything should stop him getting across, and I believe that everyone of us would dash under there to carry him over. Jacky, however, gets over safely; then, when he is on the other side, is quite cocksure over it.

Crush arrives soon after. He has been walking around the headings to see which is the best method of doing his work.

" As usual," he grunts, " them blasted afternoon blokes have left it as I've got to do a week's work afore I can get a start. Gi's a lift down on the level, will you? "

" Off the road, I'll bet," I say as we move off.

" What d'you expect? " he demands. " An' right down to the axles, all four wheels. That's the way they treats me."

Crush hitches up his trousers, spreads his legs, and flattens his back against the front of the tram. My mate collects some pieces of wooden wedges, whilst I follow the example of Crush.

" Up she goes," Crush says, and our heels dig into the floor of the roadway when we lift.

She takes some lifting, for she holds over a ton of coal, without the weight of the iron tram. As, very slowly, we lift her, my mate makes sure of the gain of each inch by packing wedges under the wheels. We use the same method at the other end, then, when the whole tram is high enough, we lean against the sides, with our feet against the tram, and push it over on to the rails. It may take half an hour or more to get the tram right, but sometimes it is done quickly, as it was in this case, and so Crush relents a little and becomes more affable.

" Sweating, see? " Crush draws his hand across his forehead and holds it in front of my eyes to show that he is speaking the truth. " And that's quite agen my principles. The doctor told me as my sweat was worth a guinea a drop. I've lost a hundred quid already. That won't pay."

Crush couples another five full trams of coal to that one, then knocks on the signal wires and the rope pulls them away. An engineer who is crouching alongside an engine about three hundred yards away will draw that coal back to a place where a bigger engine will pick up those six trams and another six from another district and will pull the lot back to where another bigger engine will take the twelve back to the main engine, which can handle twenty-four trams easily.

We pull all the smaller stones that we can see hanging down by tapping them with a long stick that is used to measure timber. Soon Crush returns with empty trams. We push one on, then start to throw stones into it. At the least sound of ripping or of dropping stones we leap back under the shelter of the standing timber. Thus we work, always tensed for the least sound, always ready for an instant leap.

We have filled that tram in twenty minutes. As stone is heavier than coal, it would hold nearly two tons. Crush kicks the sprag free; the tram runs down the slope and is coupled to the others. Another empty is taken in, filled, and run out. When we are filling the fourth, and Crush has gone to take the loaded ones to the labourers, the fireman arrives.

"What's it like now?" he asks.

"Bit quieter now," we answer; "but it's none too good. It have been all on the go."

"Let's have a look." He peers out from the shelter of the timber, says "Huh," and then steps back. "How many have you had?" he asks.

"This is the fourth," we answer truthfully.

"Huh!" he says again and walks away.

"Have he bin here?" Crush asks when he comes again.

"Yes, and gone," we answer.

"Didn't come near me," Crush complains. "He

never do at the beginning of the shift. He knows what's safe for him. Like him to say just two words wrong to me, I would that."

"He just had a look at the place and went on," I explain.

"Looked at it, did he?" Crush is being sarcastic again. "If he looked at it everything is all right. Look at the hole and the top is safe, and if he looks at the fall it's as good as clear. That's the best of being one of them officials: you can tell a bloke to do a thing even if you can't do it yourself. I'd just like one of 'em to say something out of place to me now, I would that. I'd smack 'em in the chops hard enough to knock 'em from here to the Rhondda Valley."

That would be a pretty hefty smack, as there are several miles of solid mountain between us, but I dare not doubt the ability of Crush to carry out his threat, or he may try it on me. When the fifth tram is filled, we have skirted along the sides of the fall and have cleared a distance of about four feet from the last timber. If we can get more timber up now, we will have covering to take us farther on again. I watch the hole very carefully, whilst my mate cleans a hollow down to the solid and about a yard back from the rail. When he is ready he watches whilst I do the same on my side. When I have scraped below the level of the fall, the small coal feels quite warm. I do not like that kneeling down to clear a hole, because I cannot jump back quickly from my knees. It is a quick scramble and dive to get that hollow ready, then we hurry back to load timber. They are some distance back on the side, but Crush brings his journey of trams to carry them. They are big nine-feet posts, as round as a man's body. It needs the three of us to lift them on the tram. After we have slid them off near to our working-place I touch the hatchet with a sharpening stone, then shape the timber by slicing the largest ends of two of them to a point in such a way that there is a sharp slope on the one side and a longer slope on the other. We call these sliced posts "arms"; they are to be placed upright in the hole, and the other post is to be notched to hold cross-wise on those slices.

Crush helps us to drag them into place and set them upright; it takes all the strength of the three of us. One watches the butt and sees that it slides into the hollow, while the others gasp and strain in the effort to lever the rest of the post upright. We have to be most careful, too, that we do not bang against the sides, or we may pull a fall on top of ourselves. When we have it safely in position and held with steel forks, we slide stones around to help keep it solid, and pause to regain our breath; then we measure the distance between the two arms with a measuring-tape. That will be the distance that we must notch the cross-stick, or collar.

It is too high for us to reach from the ground, so we push an empty tram forward and place scotches to hold it.

"Best to chuck some muck in it," Crush advises us wisely, "an' then you won't be so likely to slip on the iron if you gets stepping about lively. And if the tram was half-full or so it wouldn't be so hard to jump out of it."

As we reach for our shovels a small stone drops. It hits the left side of the tram, then slithers to the ground.

"Hear that little pebble?" Crush moves forward to show his light upwards. "A man's hand wouldn't be much good to him agen if he had it on top of the tram when that bit fell, or his head neither—less he was a pretty tough sort of guy, the like of me, f'r instance."

"You'd have dodged it, I suppose?" I try to banter him.

"Well, you can bet your last pint," Crush assures me, "that I'd have a hell of a good try." He taps the side of his lamp to brighten the light and glares up into the darkness just above. "What's the matter up there?" he demands of some unseen listener. "D'you think as you can frighten somebody, or what?"

As we re-start to throw into the tram, we hear a sound like calico being ripped high above our heads, then a shower of small stones falls. We jump back, listen awhile, dash in again to fill until more falls again, then leap back to the safety of the standing timber. We work that way until the tram is level full, jumping forward to grab a stone or force the shovel under the small stones, then jumping back with the burden. We are always watching and listening, and always afraid that we might not jump quick enough. We do not fear the small stones so much, for their little "taps" are part of our working life, and a small cut or a large swelling is not considered worth mentioning; but when these small ones drop, they are usually a warning that bigger ones are moving above. Even a small piece of iron stone—say about seven pounds in weight—would not improve a man's feelings if it hit him when falling from the top of that eighteen-foot hole that is in front of us.

"Keep on pinching like this," Crush observes, "and we'll have another nice little pile down. Have them mountain sheep coming down through that hole directly."

It seems that Griff, my workmate, has forgotten all about the danger of falling stones and crushing top, for he is standing behind the tram looking intently at a small flake of stone he is holding. I guess what he has found, and pause to join in the scrutiny.

In that eight-inch-wide piece of stone is embedded the print of a complete fern leaf. It is a fossil with every strand of the leaf showing plainly.

" A beauty, ain't it? " Griff asks, and I agree. Then Griff puts it carefully in his jacket pocket, because he wants to give it to his son for the school teacher.

" They'll tell him a lot of fairy tales then about coal coming from trees and these fossils proving it," Crush objected. " I can't swallow them yarns. Thousands of years ago, they says. Bah! "

" How did it come, then? " I ask him.

" How did it come? " Crush is beaten for a minute, but not for long. " Why, the landlords put it there. 'Course they did, or why should we pay 'em a tanner a ton royalty if they didn't? Sneaked over in the dark and shoved it into the mountain, that's what they did, when nobody was looking. Nice kind blokes them landlords are, you believe me."

" All right," I said, " don't get your bit of hair off with me. I don't pay 'em."

" Yes, you do," Crush insisted, " and every other blighter as works here. Comes on the cost of production it do, and a good many other things as didn't oughter. Cost of production—I brought that lot out O.K., didn't I? Think I'll stand for Parliament. By golly, there'd be some'at going on if I was to get there."

Crush is strangely silent whilst we finish the tram.

I can sense that he is thinking deeply as he stands behind me. Probably he is rehearsing his maiden speech. I go back to the third post and study the shape of the bends. We roll it over so that I can notch it in such a way that the bend will be upwards.

"I'll pop along and give that collier in the deep another tram," Crush said, "or else he'll be bawling all over the pit. I'll be back to give you a hand with that collar. Pretty big one, too: it'll take about eight of us to lift it."

After I had notched both ends at the distance apart shown by our measuring, and had trimmed the ends of the collar so that it would not touch the sides as we lifted, I went to get more help. The haulier was pulling trams of muck into the road where four labourers were at work throwing the rubbish from the trams into the sides where the coal had been extracted, and were using the larger stones to build a wall. They were working at the limit of their speed, and the stone-dust was around them in a haze. The heat was enough to check a man's breathing; the sweat was running down their faces. I noticed that the trousers of one were wet with the sweat that had soaked from inwards. The horse was clipped trace-high and was moving leisurely, but sweat was dropping from under his belly like leaking water. The fireman was pacing back and forth between the workers, his presence a silent menace to their peace of mind. The yellow light of his oil lamp contrasted with the glare of the four electric lamps that were hanging on posts.

I told him briefly that I needed help with the collar, and he grunted back: "Manage with one of the labs. We've got a lot of muck here to get rid of. Take the collier and his boy."

I knew that the collier was on piece work, as did the official, so I was doubtful if he would spare the time to help, but he was quite obliging.

" Be there in half a tick," he said; " let's just finish this tram off. It's about time for food, and it'll be fresher back with you."

He lifts a lump of coal to his knees, then shoulder high, and slides it to its place on the side of the tram. He does the same with another lump, then Jacky throws some small coal on top. The boy is obviously weary, and drops the shovelful on the edge of the tram to rest his arms after each throw. I take another shovel from the side and help to finish the tram.

" Start to mark him, Jacky," his mate says, and Jacky gladly drops the shovel, finds a lump of chalk in the semi-darkness of the stall-road, and rubs the old chalk marks from the tram sides with his cap. Then he chalks his butty's coal number at each end of the tram, and the number of trams he has filled is written in the centre panel.

" There "—the collier smooths his hand across the top of the tram—" that's enough on that one. Twenty-five hundredweight on this 'un, surely, eh? That's earned another two bob. Bring your grub, Jacky."

Crush is already waiting; with him is another repairer named Ned and his mate.

" Ned was coming back to you," Crush explains, " as he wanted you to put a patch on his arm. S'pose as you've got plenty of that oil handy, eh? "

Crush always describes iodine as " oil ". He is convinced that one of the great pleasures of my life is that of rubbing iodine around the cuts that my mates get. I examine the arm of the repairer whom Crush has suggested that I " patch ". The skin has been lifted from above the wrist almost to the elbow. The stone that lifted it must have been as sharp as a razor. I clean around the edges with cotton wool, try to replace the skin, and iodine around the edges. Crush shows light and gives encouragement.

" He've got plenty of that oil if you likes it," he states, " and he enjoys dabbing it on."

I place a large wound-dressing over the injury, then wind the attached bandage around the arm. To make a better covering I follow it with a two-inch bandage that I spiral up the limb. Crush is pleased with the effect.

" Good as ever now," he says, " and you can give us a lift with the collar now? "

" Better rest that arm for awhile," I suggest.

" I'll go gently, but I'll give you a lift afore we goes," the repairer agrees.

We could have done with more help, but as we are usually short-handed, we do the best we can. Between us we lift that collar on to the tram, then roll it so that it is in the right shape for lifting. We arrange our lamps so that they will show us light, then we listen awhile to note if anything above is moving.

" Seems fairly quiet for a bit," Griff says. " Let's stick him in place."

Two of the tallest stand on each side, using the stones as a platform. The others clamber on top of the tram and bend to get their arms around the post. There is a violent hurry to lift before anything falls:—

" Up she goes, my lads. That end first. Hold this end steady. Up—up—up. Resting on the arm, is she? In place—what?—bit more that way. She goes. Right. Now then, two of you hold that end steady, the others help us. Now then—up—up. Heads under it. Don't let her slip, lads, or she'll cripple a couple of us. Up agen. Ah! she's there. Fit's like a glove. Take a blow, boys. What's a bit of a post to men like us?"

She is in position, and we cock our heads from side to side as we view her. Must tap that arm with the sledge a bit to get all square, then it'll do.

"D'you guys know as it's near four o'clock," Crush demands, "and it's past grub time? Better have it now, unless you're going to take it home and have warm tea with it. I'm having mine. If I can't have time for grub, I can't have time to work, 'cos it's only grub we're working for."

We sit with our backs against the hard stones, after we have gone some distance back to a safer place, and no armchair ever felt more welcome and comfortable. In the outside world another day is dawning, and millions must be sleeping in their beds, but we are going to have our dinner. We swill the dust down our heated throats, we open our food-boxes and use the paper to shade our lamp-glasses. We put the lamps in a circle around our feet so that the rats dare not creep too close. We can hear them running about amongst the stones at our back, and occasionally a moving shadow can be seen outside the range of the lamps. One of us throws a stone at the shadow, which disappears. From the distance sound the squeals of rats when they fight over some scrap of food they have found.

"Sounds like hosses moving about, they do," Griff says. "I wonder what they finds to live on? Thousands of 'em here, and nothing much to eat except the bit of feed as the hosses drops from their nose-bags."

We have put our jackets on so that our heated

bodies shall not be chilled. Our singlets cling like an extra skin to our chests. We have made the boy sit in the middle of the line of men; we feel we must shelter him from the danger of falls and the rats.

Griff knocks the top and bottom of his food-tin together as a sign that he has finished. He looks at his watch.

" Got another four minutes to go yet," he says —" time to stretch me legs. Lord! Couldn't I sleep? I wouldn't want no spring mattress—just leave me alone on these stones."

" Here's one as would keep you company," said Crush. He was reading an advertisement on a piece of paper. " Do you feel sleepy after meals? " " Sleepy after meals, by God ! " he says. " It's in by here they oughter be. That fireman 'ud see as they didn't have chance to be sleepy after meals. What do you say, Jacky? "

Jacky made no reply, unless a faint snore could be counted as one. A piece of bread and butter was still clasped in his right hand; the white of some unchewed food showed inside his lips. He was sleeping peacefully on his bed of small coal, with two stones for a pillow.

" He's gone, poor kid ! " Griff whispered. " Let's keep quiet, so's he can have another minute."

So, half-risen to return to our work, we waited, so that the boy should have another minute of rest. It is against the law to sleep underground, but physical exhaustion sometimes ignores man-made laws. After a short wait Crush got up reluctantly.

" Sorry, kid," he muttered, " but I've got a heap of work to do before half-past six. I—what the hell is coming off now? "

For Jacky was disturbed in startling fashion. A

sharp crack sounded from our working-place, then a clatter of falling stones, and a loud crash. Everything was silent afterwards, and we turned our heads away to avoid the dense fog of dust that moved back past us. Jacky had come to his feet in one leap, and in that action a startled cry of "Mam" had issued from his lips.

"It's all right, boy," Crush assured him—"just a bit more of a fall for these blokes to clear. But it was a good job as we let you have that extra minute, or some of us would ha' been pretty flat by now, so I'm thinking."

After the dust has thinned a little, we venture forward to see what has happened. The smaller stones are not such a problem, but as we come near the tram, we find our way barred by a huge stone.

"That upper top have come," Griff is the first to decide: "and yet it sounded as if it was strong when we tapped it—well, it sounded decent, anyway."

"My stars!"—Crush looks at the stone—"a good three ton in that sod, I'll bet. Think of that lot dropping on top of a man, hey? He wouldn't have enough breath left to squawk."

I feel sick when I look at that stone. It has crashed down on the exact spot where I was shovelling. The weight of the stone and the height of its fall have forced the ground up around where it is lying. Ugh! If it had happened a minute later it would probably have caught me, and escape would have been impossible. Then Griff discovers something more.

"Look at our collar," he calls—"that stone have split it. Aye, it have, right across. We'll have to change that 'un agen. As if we hadn't got enough on our plate."

The collier promises to tell the fireman of this new happening; and Crush gives us another tram. It keeps quiet around us, and we hurry to fill all the small stones in this tram. Then I get the sledge, and we hammer away at the big stone. Time is getting short, and we are warming up properly now. Our heads are streaming with sweat; it squelches in our boots when we move; it runs into the cuts and scratches on my arms and makes them burn; I can feel the perspiration running down my legs.

We cannot get the real grain of that stone. We walk around it and tap it from different angles, but find no mark of a split. The blows from our heavy sledges leave only white marks on its side—it will not break. The fireman arrives and is impatient.

" How did this fall? " he asks.

I think that a fool question to put to weary men and reply:—

" Downwards, and damned hard at that."

" I know. I know that," he splutters; " but couldn't you get something under it? "

" It was too high," I said; " and we was getting timber under the bottom piece. Had the collar in place, but this stone cracked it when it fell. Was a good big collar too—a Welsh oak."

" Cracked him? " He is incredulous. " A new collar, too? "

" Have a look, then," I suggest, and he does so.

" Lord! it must have dropped with a crash," he says, when he steps down again near me.

" It did that," Griff agrees, " and without an ounce of warning. It was down in a flash."

We pound away at the stone, whilst the fireman fumes and mutters with impatience as each blow fails to break it. Soon he loses patience.

" Chip a hole on the top of that stone," he suggests, " and fetch some of your powder. Three pills 'ul be enough, don't you think? It'll be quicker than pounding, and I'll go to get the battery."

It took some time to chip a hollow in that solid stone, and by the time we had done that, and I had fetched powder, whilst Griff had collected a shovelful of clay, the fireman had returned with the battery and fuse. We pushed two pills of powder into the hollow, worked a detonator into the other and placed it alongside, then covered the lot with clay, except the ends of the fuse-wires. While we were at this, the fireman had climbed on the side and was lowering the flame of his oil lamp until it was no bigger than a pin-point. No blue cap of gas collected on the flame, however, and so he lifted the flame and stepped down alongside us.

" Pretty good flow of air just here," he stated. " Are you ready? "

" All set," I reply. " I'll go along the level, and Griff will walk up the rise heading. You'll watch this way."

He uncoiled the fuse and it was connected to the wires; then he went around the corner and started to crouch against the side preparatory to connecting to the battery.

" I'm firing as soon as I've connected," he warned us as we hurried away.

" All right," we answered over our shoulders; " we'll be far enough."

When he fired, the workings all around seemed to shake. The roar filled our heads and deadened our hearing. Dust passed us in a thick, slow-moving cloud that appeared to be solid. After the dust came the powder-smoke, burning and blue to the eyes, and acrid to the nose; it made me cough and gasp before the relief of sneezing came.

Crush was waiting back there; he had stopped his rope.

" That gave him a heck of a clout, by the sound of it," he said. " That John Noble is a hard hitter if you gives him a chance."

" Aye," I answered, " and I'm waiting for the smoke to clear a bit before I starts. Might be some loose bits about."

" I can go on," Crush said, and knocked three times on the wires.

He had not travelled a dozen yards when he was nearly caught. He was standing on the coupling between the third and fourth tram when a wooden lagging between two pairs of timber suddenly snapped, and one of the sharp pieces pointed right across that roadway. Crush had ducked his head at the first sound, thinking it was only the usual crackling that heralds the breaking of timber. This piece of broken lagging, however, grazed one tram, then swung right across between the two trams where Crush was riding. Probably it was his instinctive swaying away of his body that saved him from being transfixed, and it was fortunate that there was enough space for him to tumble away on the other side of the trams. I heard his shout, and jumped to tangle the wires. The journey of trams stopped almost immediately. Crush got to his feet tenderly, but his language was quite robust.

" Why didn't you sidestep that one? " I asked.

" I did, mun," he argued. " Didn't you notice how I swayed me body to one side? If it had been an ordinary man, it would have killed him."

That " ordinary man " has been a favourite joke with Crush since the time when a small stone fell on an Irishman who was working with me. He was a weakly Irishman, and under-sized, but he had his share of conceit.

"Begod," he declared, rubbing that part of his head where the stone had tapped, "if that stone had hit an ordinary man it would have killed him, it would that."

Crush is no squealer. He does not murmur when I pull his singlet up past where the skin has been stripped from his side. I examine the ribs, and he insists that none of them are broken. I paint around the wound with iodine, and as I do so he grins at me.

"Enjoying yourself, ain't you?" he says. "But go on, you; I'll have a chance to get me own back some day, and then I'll make you hop."

I am doubtful about his ribs, and insist on bandaging him tightly and following with two triangulars—just in case.

"Could think I was in a blasted strait-jacket." Crush straightens himself with difficulty.

"They'll support you, anyway," I said, "and I'll take the trams along for you, if you like. Better for you to sit down for awhile until we goes out."

"I'll manage all right," he insists. "You don't think as I'm a blasted kid, do you? 'Sides, you blokes have got a gutsful to finish your own job. There's another two trams there yet."

There were, but the powder had smashed the stone small enough for us to handle the parts, and we had cleared them by the time the lights from above started to move down towards us. As the lights parted from one another by coming in, so now on the outward journey they met again as they passed each branching passage, until the mains were a constant stream of hurrying shadows that carried lamps. Crush swore a great deal whilst he was putting his shirt on, and said, " Well, I reckon I've earned my two pun ten this week agen, if it's only for the bumps as I've had."

After about half-an-hour's tramping we saw a distant circle of daylight When we got outside the arching, the brightness dazzled our eyes. We waited in a crowd below the boilers with that " Danger " signal behind us. The mist that heralds a fine day rose from the brook alongside, and the sky showed blue and clear above the brown sides of the encircling mountain.

The officials stood in a row, keeping the men behind them until the exact moment. The men muttered and shuffled about impatiently behind them. They switched off their lamps because they were now unnecessary. At the lamp-room the attendants slid the shutters back ready and turned the check-board around so that they should hand the checks out swiftly. Crush pulled up his moleskin trousers and stretched his legs.

" How d'you feel now? " I asked him.

" There," Crush stated definitely, " that's the end of another week. I'll have a look at a game this afternoon, and a couple of drinks after. Then we'll have a good dinner on Sunday, and I'll have a good sleep. When I wakes up it'll be near time to come to work, and another week-end 'ull be over."

" Not much of a prospect, is it? " I asked.

" Course it ain't," he agreed; " but what else is there? "

" Not much, I know," I answered; " yet it could be a lot different."

" Aye, I knows as it ought to be," Crush said, " but——"

At the nearest of the boilers a stoker walked across to a lever. He lifted his hand up, then pulled his watch from his waistcoat pocket.

"He's behind, as usual," one of the waiting men complained.

Suddenly the stoker jerked his hand sharply downward and the hooter blared out. The crowd of men surged forward.

"There she goes," Crush shouted; "let's hop it."

And we went, swiftly.